What people are sayi

Healthy Planet

The danger is not just more pandemics like Coronavirus. The solution is not just a greener "recovery". *Healthy Planet* shows that the fundamental disease is the wrong relationship between people and planet. Climate breakdown and biodiversity loss are far more dangerous than COVID-19. Fred Hageneder does not pull punches about the real state of the world but he charts how we can heal its wounds.
Sandy Irvine, MSc Activist and author, Newcastle-upon-Tyne, UK

Healthy Planet is a very timely publication. Fred Hageneder captures the true wonders and complexities of Planet Earth and all its inhabitants. He knows how much damage we humans have caused, but provides many practical suggestions for creating positive change. His message is clear and will inspire many people to take action before it is too late: together we can, we must and we will save our planet.
Jane Goodall, PhD, DBE Founder of the Jane Goodall Institute and UN Messenger of Peace

A practical and beautiful compendium on the state of the Earth and what we can do to save it.
Prof. Dr. Ernst Ulrich von Weizsaecker Honorary President of the Club of Rome

This is a uniquely critical and simultaneously optimistic book about the future of this planet!
Dr. Helen Kopnina The Hague University of Applied Sciences, The Netherlands

The comprehensive scope and sheer amount of information gathered here is impressive.
Prof. Eileen Crist Virginia Tech, USA

Healthy Planet is a uniquely clear and comprehensive guide to the most important issue of our times, and includes invaluable suggestions for what to actually do about it.
Patrick Curry, PhD Author of *Ecological Ethics*

With an eye on future generations this book offers practical advice for the wise stewardship of this rare and precious planet—the only home we know. Fred Hageneder has crafted a hopeful and empowering *cri-de-cœur* for the guardianship of Gaia.
Dr. Kevan Manwaring University of Leicester, UK

Healthy Planet

Global meltdown or global healing

Healthy Planet

Global meltdown or global healing

Fred Hageneder

Winchester, UK
Washington, USA

JOHN HUNT PUBLISHING

First published by Moon Books, 2022
Moon Books is an imprint of John Hunt Publishing Ltd., No. 3 East Street, Alresford
Hampshire SO24 9EE, UK
office@jhpbooks.net
www.johnhuntpublishing.com
www.moon-books.net

For distributor details and how to order please visit the 'Ordering' section on our website.

ISBN: 978 1 78904 830 8
978 1 78904 831 5 (ebook)
Library of Congress Control Number: 2021938947

A CIP catalogue record for this book is available from the British Library.

Design: Stuart Davies

UK: Printed and bound by CPI Group (UK) Ltd, Croydon, CR0 4YY
Printed in North America by CPI GPS partners

Contents

Previous books by the author

Yew (Botanical Series). London 2013: Reaktion Books, ISBN 9781780231891.

Yew: A History. Stroud 2007/2011: The History Press, ISBN 9780752459455.

The Living Wisdom of Trees: A Guide to the Natural History, Symbolism and Healing Power of Trees. London 2005/2020: Duncan Baird, ISBN 9781786783332. Published in the USA as *The Meaning of Trees*. San Francisco 2005: Spectacle, ISBN 9780811848985.

The Heritage of Trees: History, Culture and Symbolism. Edinburgh 2001: Floris, ISBN 0863153593.

The Spirit of Trees: Science, Symbiosis and Inspiration. Edinburgh 2000/2006/2017: Floris, ISBN 9781782504481. New York: Continuum, ISBN 9780826417633.

For Charlie, Blossom, and Myla,
Tula-Rose, Leala, and Quinn.

And everyone who chooses
this beautiful planet.

"Healthy people and a healthy planet is part and parcel of the same continuum."
– Inger Andersen, Executive Director UN Environment Agency, 2020[1]

"Please join me in demanding a healthy and resilient future for people and planet alike."
– António Guterres, UN Secretary General, 2020[2]

"Urgent and inclusive action is needed to achieve a healthy planet with healthy people."
– UN *Global Environment Outlook*, 2019[3]

I can lose my hands, and still live.
I can lose my legs, and still live.
I can lose my eyes, and still live.
I can lose my hair, eyebrows, nose, arms,
and many other things, and still live.

But if I lose the air I die.
If I lose the sun I die.
If I lose the Earth I die.
If I lose the water I die.
If I lose the plants and animals I die.
All of these things are more a part of me,
more essential to my every breath,
than is my so-called body.

What is my real body?
– Jack D. Forbes, First Nation activist, writer and scholar, 1979[4]

Introduction

The Earth is an interconnected web of life. Our species is part of this planetary community, always has been, and will be till the end. The health of the planet is our health.

Already the Greek physician Hippocrates (c. 460 BCE-c. 370 BCE) knew that public health depended on a clean environment.[5] Some two millennia later, this is beginning to be understood at government level and in the UN. Human health, animal health, plant health, and ecosystem health are inextricably linked. We cannot separate ourselves from all life on this planet; no single species can exist on its own. We are part of the very ecosphere we live in: what assaults one also harms the other.

When we poison our agricultural fields we also poison the insects, the birds, the animals, and ourselves. When we dump toxic materials in the sea we expose all sea life to suffering and harm, and just like the fish, the sea otters, the seals, and the orcas, humans end up with liver and kidney failure, immune system breakdown and cancer. Whatever substances we blow into the air, will also be absorbed by our own lungs. By now, microplastics have contaminated the entire planet, and we find particles in the Arctic as well as in our own cells.

One planet. One health.

Humanity is totally dependent on the very ecosphere it is degrading. But we live and consume clearly above our planetary means, we exhaust the "carrying capacity" of the Earth. Population pressure, overconsumption and a ceaseless production of waste bring the life support system of the planet to its knees. The water pathways and the climate are in serious disarray. In arrogance and hubris we have piled ecocide upon ecocide until biodiversity loss turned into a global crisis: the sixth mass extinction in Earth's history is underway. It's not the first one, but the first one that is anthropogenic, man-made. And the speed of this biodiversity loss is by orders of magnitude faster than any in natural history.

And yet, even now that we read the unmistakable signs that our house is burning (*eco* means "home, dwelling"), we cannot seem to stop. In the rush for infinite "economic growth" we rush towards an early grave, not just at the personal but also at species level. By wanting everything for ourselves we jeopardize the well-being of future generations, not just human ones but of all our fellow creatures on this planet.

The next decade will be the most decisive in the entire history of humanity. Will we irreversibly destroy Earth's benevolent climate? Will we fail to counteract the sixth mass extinction? Will we annihilate our own race? Why are governments and leaders acting so slowly? And what can individuals do?

We have a very few years left to profoundly shift our attitude, presumptions, and direction. This book is trying to provide some inspiration and useful insights to help with this shift. We have the approaches and solutions (and the necessary sustainable technologies) to deal with the mess we've made. But we need to wake up and apply them.

This book starts with a holistic picture of the planet, *Part I: Living Earth* shows how an originally healthy planet works, how its ecosystems evolved and how they interact globally. This is a solid but accessible introduction to Gaia Theory and Earth system sciences.

Part II: Global Disruption looks at how we continue to disrupt and degrade the planetary life support systems. For each area (climate disruption, mass extinction, pollution, etc.) there are both global maxims for action as well as helpful suggestions as to what each one of us can do (admittedly, personal efforts have a ridiculously small impact, but they do add up if many people begin to take responsibility).

But the solution to our problems cannot be found within the old ways of thinking that created them. Hence *Part III: The Human Interface* tackles the deeply ingrained paradigms that bind us to our deluded dance of destruction. Humanity's collective mindsets, the underlying philosophies, our inherited values have to change.

We need to re-think and re-feel, to empathize with other creatures, and to honor life again in real and meaningful ways.

A beautiful healthy planet is possible again. Let us make the right choices and take all the necessary steps.

Fred Hageneder

Part I

Living Earth

Planetary Life Support Systems and their Interdependence

Complex regulations and mutual interdependence link together every animal and vegetable form, with the ever-changing Earth which supports them, into one grand organic whole.
– Alfred Russel Wallace, co-progenitor with Charles Darwin of the theory of evolution, 1876[1]

Chapter 1

Foundations

The Earth is not a ball of rock floating through space with some life forms dotted about on its surface. Neither did life occur on a planet that accidentally finds itself in a "habitable zone" (not too close and not too far from a sun), nor did biological life colonize some "niches" to dwell in. A very different picture has emerged over the last few decades that reveals how much the entirety of life (the biosphere) *actively* maintains the viable conditions on Earth. "The biosphere is not simply *in* a habitable zone but also *makes* a habitable zone," say Earth scientists Eileen Crist and Bruce H. Rinker.[1]

For the 3,800 million years of life's existence, the living *(biotic)* and non-living *(abiotic)* domains have amalgamized so profoundly "as to form a biogeochemical entity that behaves as a self-regulating system." (Crist and Rinker)[2] In other words, *the organisms shape the conditions of their environment to their advantage*. And keep it that way. (These are long-term effects of *planetary* evolution and don't collide with the notion that individual species have to adapt to circumstances within the timeline of their own becoming.) To make this less abstract:

- Planet Earth when it was young would have lost its water if it wasn't for the work of myriads of bacteria. Their metabolism released free oxygen or certain sulfur compounds that were able to bind the light hydrogen atoms. Thus the microorganisms prevented the escape of hydrogen into outer space. *Without life there would be no water on Earth.*
- Land plants and therefore animals (including humans) depend on fertile soil; which would not exist without

bacteria gaining and preparing mineral nutrients from the bedrock.

- Did you know that 99 percent of the atmosphere comes from living beings? A fifth of the air is oxygen exhaled by photosynthesizing plants and algae, and four-fifths is nitrogen purified and supplied by bacteria. Without life, Earth's atmosphere would be a mix of toxic gases, and boiling hot at that. As it is on our living Earth, the elements of the air have only recently existed as parts of living cells.

Note that bacteria play a crucial role in all of the above points; we'll get back to them later. The last point mentions surface temperature and that is indeed the prime example for the self-regulation of a living planet.

Amidst the vast range of temperatures possible in the physical world—from total zero to millions of degrees—the window suitable for biological life is extremely narrow: zero to 122 degrees Fahrenheit (50 degrees Celsius), with few exceptions, like thermophilic microorganisms living in deep sea vents at much higher temperatures. Because proteins coagulate at 107.6 Fahrenheit (42°C) and hypothermia threatens below 95 Fahrenheit (35°C) body temperature, humans and other mammals have an even smaller window. Plants photosynthesize best at about 73 Fahrenheit (23°C) which sets a mark for the optimum temperature for land-based ecosystems. The optimum for the oceans is 50 Fahrenheit (10°C) or less, which enables the most efficient mixing of surface waters with bottom waters through *convection*, bringing nutrients to the surface and oxygen and CO_2 to lower strata. Putting land and sea together *the optimum global average temperature is about 59 Fahrenheit (15°C). This is the ideal working temperature for planet Earth.*

When the planet was young its own heat was far too high for living organisms. As it cooled on the magmatic inside, the greenhouse gases in the atmosphere (mostly volcanic CO_2) still kept the surface too hot. But over hundreds of millions of years, the tireless photosynthetic work of microorganisms and early plants changed the atmosphere and with it the global temperatures to what we (still) know today. But the amazing thing is this:

Astrophysicists tell us that *since life appeared 3.8 billion years ago the sun's energy output has increased 25 percent*. But as we know from geology, paleontology and other Earth sciences, life has been present in an unbroken line since its very beginnings, which means that average surface temperature has always been about 59 Fahrenheit (15°C).

The discovery in the early 1970s of the planet's obvious ability of self-regulating its temperature led to the new academic branch of Earth system science. The living planet is seen as an interconnected web of eco-systems, with inherent abilities of self-regulation and self-generation. It has been (re-)named *Gaia,* after the primordial goddess of the Earth in ancient Greece. Gaia is more than a synonym for the biosphere. **Gaia is the entirety of the material Earth and all biota (living organisms) on it.** This "grand organic whole" is capable of self-regulating the temperature and the chemical and physical composition of the planetary surface in order to sustain comfortable conditions for life. This requires active processes of automatic feedback, the energy for this is supplied by sunlight.[3]

Gaia Theory and Earth system science

In 1973, the British scientist James Lovelock, who had been working on the NASA Mars program for years, published his first paper outlining planet Earth as a complex superorganism. The "Gaia hypothesis" didn't have an easy start because it fundamentally integrates diverse fields of science such as biology, geology, oceanography, paleontology, and mineralogy, among others, into one single approach of systems theory. In an age where Western science is deeply "reductionist", splitting up into ever smaller compartments (biology alone has over thirty), Lovelock's holistic approach to understanding planet Earth was nothing less than a provocation. Especially for the neo-Darwinists who countered that a planet cannot "evolve" like living organisms.

However, the Gaia approach sparked the development of what today is called Earth system science, which brings together more than twenty disciplines. Around the turn of the millennium, Gaia *hypothesis* matured into Gaia *Theory* and is now widely accepted. Particularly noteworthy is the presence of Gaian principles in the swiftly developing climate science. As late as 2012, the computer models of climatologists were rightfully criticized for not adequately incorporating the impacts of the biota, namely forests like the Amazon, on global climate.[4] Since then, under the pressure of the harbingers of climate disruption, climatology has adopted many of the Gaian views about the interconnected Earth system. Modern climatology can not be separated from the Gaian perspective anymore. It was high time, but is also an irony: it's the way it is in kindergarten: only by destroying something, Man can see how it works.

But it is not just temperature; many physical properties of the Earth system need careful balancing:

- global temperatures, weather and climate;
- the salt levels of the oceans;

- the oxygen content of the atmosphere;
- the (chemical) reduction potential, especially of the atmospheric gases;
- the electricity of the air;
- the acidity of air, water and soil;
- the availability of water on the continents;
- the dispersal of mineral nutrients;
- the strength of cosmic radiation.

The network of ecosystems and their biota is closely interlinked (like the organs in our bodies) and the planetary "metabolism" of matter and energy facilitates Gaia's active regulation of the above parameters. It is no wonder then that Gaia has been called a *superorganism*. Using this term to describe Gaia is rather controversial in science because, strictly speaking, an "organism" in biology is by definition able to reproduce, and over generations its species can evolve through genetic inheritance and adaptation. It is true that planets don't create offspring together but the Earth surely does evolve (see next section).

In layperson's terms the comparison is fair and square anyway. Just as the cells in our body form organs and tissues which communicate and cooperate with each other to make up the metabolism of our body, so do the animal and plant species form ecosystems which communicate and cooperate with each other to make up the metabolism of planet Earth. Ant and bee colonies are being described as superorganisms. The human body is, and, in a wider view, a human society is a superorganism. And so is Gaia. In all of these, the elements work together and form a whole which is more than the sum of its parts. Both the Earth and our bodies are populated by myriads of bacteria whose constant activities enable the organism to be alive in the first place.

And haven't Indigenous peoples always held an organic

view of Earth, calling rivers and waterways the bloodstream of Mother Earth, the wind her breath, and the rocks her bones? And these are exactly *the three domains of Gaia: the sea, the atmosphere and the crustal rock.*

The sea

Life comes from the sea. And the sea is still the richest ecosystem with the highest biodiversity. And everywhere water enables all biota to live. Warming in the sun, the surface strata of the oceans generate vast amounts of water vapor which condenses to clouds to bring life-giving rain to the creatures on the continents (although the bigger part rains back into the ocean). But however damp the air, clouds don't seed themselves. It is the vast fields of algae which emit certain compounds that act as condensation nuclei for cloud formation. Furthermore, the oceans are major players in the regulation of the global climate: the sea absorbs CO_2 from the air and is the world's biggest carbon storage, and the white sea clouds have a high *albedo* (reflection) which reflects solar energy, thereby keeping the planet cool (see Chapter 5).

The air

Its components created by living beings themselves, the atmosphere is a perfect matrix for exchange between life forms and between ecosystems. Gases, liquids and solids can be shared and transported across this domain. Thus, the non-aquatic biota can find nourishment and also release their waste products. Also, just like the outer bark of a tree which protects the living tissues beneath, the layers of the atmosphere protect the living Earth from harmful influences from outer space. For example, the ozone layer captures 97-99 percent of all ultraviolet radiation. The atmosphere is a dynamic but delicate, thin layer "constantly being repaired and made whole by life itself," says Australian ecologist Tim Flannery.[5]

The rock

The mineral world provides foundation, protection and nutrition for biological life. Microorganisms greatly increase rock weathering; the breaking down of basalt rock, for example, happens a thousand times faster with the help of microorganisms than in sterile rock.[6] We are used to thinking of the transformation of the rocks of the Earth's crust as being caused by volcanoes and other geological forces, but 75 percent of the energy used to transform rocks worldwide is provided by living things such as plants, lichens and, especially, bacteria. Their work on the stone structures is three times greater in effect than that of the world's volcanoes combined.[7] Microorganisms reach deep into the rock strata and break it down with the acids they discard. Some of the minerals thus released are further processed to organic compounds which plants can absorb as nutrients. Thus *microorganisms create the foundation for fertile soil, one of the true powerhouses of life*. Some of the nutrients are washed out of the soil by rainwater and find their way into the water cycle, becoming available for aquatic creatures.

The forests

The forests hold a special position. They are not primeval abiotic domains like the physical expanses of hydrosphere, atmosphere and lithosphere (sea, air, and rock). They are complex ecosystem habitats for myriads of living organisms; in fact, apart from the sea, forests have the highest biodiversity, the highest levels of biomass (leaves, humus), and the strongest impact on regional and global climate. But to begin with, the land was barren...

When life left the ocean, amphibians could only stay close to the shore and tentatively spread along the river courses. But for life to colonize the landmasses of the continents it was essential to ensure a sufficient water supply at any distance from the ocean. Life on land required a method to transport moisture inland from the sea. Rain clouds are a good start, but they shed their water cargo after a maximum distance of about 600km.[8] How could life venture further inland? The solution was a biological one: the evolution of forest, a continuous surface cover consisting of tall plants (trees) closely interacting with all other organisms to let rich ecological communities grow. *Forests are responsible both for the initial accumulation of water on continents in the geological past and for the stable maintenance of the accumulated water stores ever since.*

The underworld

The Earth's crust is the (more or less) solid shell on an interior which is hotter and in constant flux. The layer beneath the crust is called Earth's mantle. It consists of silicate rock which is predominantly solid but in geological time behaves as a viscous fluid. The crust is divided into a number of plates which slowly move with or against each other. Side effects of these plate movements are earthquakes and volcanic belts as well as mountain chains.

In the "convergent" zones on the sea floor where one plate dives beneath the other (subduction), basalt and sedimentary layers are being returned into the Earth. Here, basalt rocks sink into depths of 250-400 miles (400-650 kilometers) where pressure (from the weight of the rocks above them) and heat (from nuclear processes deeper inside the Earth) reforge them. Eventually, in the "divergent" zones they will reappear as fresh basalt of volcanic activity.[9] If an oceanic and a continental plate meet, however, the latter is pushed up (because the continental ones are lighter) and mountains are formed. At the lower edge new continental granite is formed. Without this process, the continents would disappear completely over several tens of millions of years due to weathering.

During its formation in the divergent zones, the seabed basalt is strongly interspersed with water. This makes it flexible enough for subduction much later on. The metamorphosis of organic limestone deposits also creates an additional "lubricant". Since these processes which later help to produce the continents require large amounts of water, and because without life on Earth there would be no limestone deposits or water, we can say that life (together with Earth's internal heat) has always contributed to the formation of continents. The circular Gaian dynamic is: **no life, no water > no water, no plate tectonics > no plate tectonics, no life.**[10]

And by driving volcanic activity and continent formation, plate tectonics also release carbon to the atmosphere, thereby preventing the Earth from entering a permanent frozen state.

The whole Earth system including its mighty geological processes is increasingly recognized as intrinsic to life processes. The science journalist Richard Monastersky says in *New Scientist*: "It is now clear that the separate regions (crust, mantle and core) are engaged in a multichanneled conversation. Across major boundaries and thousands of miles, these sections exert profound effects on one another." In the same article, seismologist Don Anderson says, "you have to treat the Earth as a system; you can't just look at a part of it." And evolution biologist Elisabet Sahtouris concludes that "we can no longer consider the biosphere alone as a meaningful entity, but must speak of the whole Earth, from innermost core to the magnetic fields surrounding it, as one systemic entity."[11]

The salt of the Earth

Salt released through the weathering of rocks is transported by rivers and accumulates in the ocean. The riverine amounts are minute (hence we don't taste salt in freshwater); it takes about 60 million years for the rivers of the world to accumulate an "ocean's worth" of salt. Marine salinity varies between 3.1 and 3.8 percent, averaging 3.4 percent (i.e. 100 grams of ocean water completely evaporated leave 3.4 grams of salt). 90 percent of sea salt is sodium (Na+) and chlorine (Cl^-), other elements are sulfate (SO_4^{2-}), magnesium (Mg^{2+}), calcium (Ca^{2+}), and potassium (K^+).

Living cells control their interior salinity with intricate ion pumps in their membranes. They need to maintain their inner osmotic pressure in relation to their surroundings, and an inner electric potential favorable to their metabolic processes. Marine salinity at 3.4 percent is just perfect for life. The maximum chemical saturation of sodium and chlorine is ten times higher, and should marine salinity exceed 5 percent the membranes of cells would be torn to shreds, and, following the demise of plankton, all life in the sea would die. Geological analysis of the sediment rocks has shown that the salt concentrations in the oceans have not changed over the last 570 million years, and we also know from fossil records that life has prevailed continuously in the oceans. So where does all the salt go? What regulates ocean salinity so precisely?

An obvious answer is plate tectonics. The huge amounts of water absorbed by basalt and being melted back into the upper region of the Earth's mantle are saltwater of course. But all attempts to model the steady salinity of oceans solely "on the basis of chemistry and physics have universally failed."[12] But there is another dynamic, and once again it involves the biota, and the formation of salt plains in lagoons and sea basins. In warmer regions the water evaporates and leaves a layer of salt. Thick mats of certain bacteria create organic films over the salt which are not water-soluble, hence the returning tide can not dissolve the salt deposit. Vast evaporite deposits of salt rock (halite) exist in the United States, Canada, Pakistan, and the United Kingdom. Furthermore, salts are also incorporated in the shells or corpses of marine microorganisms and find their way to the sea floor sediments. And algae release chloromethane into the atmosphere (compare "The sulfur cycle", Chapter 3).

But a considerable amount of desalinization happens in the sediment sludge. The small size of bacteria leads us to misjudge their ecological significance. First, there is their density, "just one milliliter of sediment can contain up to 100 million salt-pumping bacteria." Secondly, their huge collective surface area: although bacteria make up just 10-40 percent of the biomass in sea water, "because of their high surface to volume ratio they represent 70-90 percent of the biologically active surface area." (Hinkle)[13]

Flannery sums up the situation of life on Earth: "We can think of Earth's rocky crust as a huge holdfast, like the lower shell of an oyster, which life has formed to anchor itself. And if we imagine the rocks as life's holdfast, then we can think of the atmosphere as a silken cocoon, woven by life for its own protection and nourishment."[14] Thus life creates a space for itself between the deadly (radioactive) fire of the planet's interior and the fatal cosmic rays.

Chapter 2

Origins

Cosmic

Looking at the origins of our planet in the macrocosm helps us to understand the Earth's relation to the universe and its role as a self-organizing and evolving entity.

According to the Big Bang theory the observable universe came into existence some 15 billion years ago in an instant explosion from an initial state of very high density and high temperature. Energy and matter as well as space and time appeared practically out of nowhere (or another dimension) in an instant of creation. In the next stage the primordial energy condensed into electrons, protons and neutrons, and with further cooling, these particles coalesced into hydrogen atoms.

The Big Bang produced none of the heavier elements. They were forged much later on when stars were born. Their gravity began to fuse hydrogen nuclei together into helium. This process releases vast amounts of energy, some of which is visible light. The universe lit up with their brilliance and for the first few billion years no other elements came into being. But as the stars aged, pressure and heat grew in the largest ones and heavier elements were formed. Carbon atoms were created when groups of three helium nuclei fused. As the stars aged even more, denser elements such as sodium, magnesium, oxygen, and iron were born from the fusion of carbon. Only the very largest stars created even heavier elements than iron.

Eventually, stars die by first collapsing and then exploding. These explosions (supernovae) sent vast amounts of hydrogen and the other early elements—by now also including sulfur and phosphorus—into interstellar space. Some of the elemental clouds that result from this slowly formed new stars, giving rise to new supernova explosions and more newly synthesized elements. About 4.5 billion years ago, our solar system formed from the gravitational collapse of a giant interstellar molecular cloud. Everything on Earth is made from hydrogen atoms which might be as old as the universe, and from heavier elements which were forged in (often more than one generation of) dying stars. *Rocks, water, air, trees, humans, birds, dolphins, fungi: we are all made of stardust.*

Planetary

From the start, our home planet orbits the sun at just the right distance for benevolent levels of heat and radiation. It has just the right size and therefore gravity to hold the ocean and the protective atmosphere. The configuration and sizes (masses) of the other planets in the solar system are so well tuned that their gravitational forces stabilize the Earth's orbit around the sun; had any of the masses of the other planets been slightly different, Earth's orbit could suffer from enormous irregularity.

Yet another fortunate event occurred when a Mars-sized body hit the young Earth and split off what was to become our moon. The moon is critically important for the Earth and her multitude of life because without the moon's additional gravitational field the Earth axis would wobble chaotically. What's more, our position in the galaxy is also perfectly tuned: the outer reaches of one of the spiral arms of the galaxy are safe from the sterilizing gamma rays that emanate when supermassive stars collapse in the center of the Milky Way.

Was it pure luck that all the right conditions lined up like this, with each of them against astronomically high odds? Was matter waiting for the right conditions to bring forth an evolving, self-regulating planet hosting a multitude of life forms? In the words of ecologist Stephan Harding, "matter ached to experience itself unfolding into the fullness of the living state."[1]

When the Earth was young it was a very different planet to

the one we know today. The outpourings of volcanic activity caused chemical reactions with water, engaging the highly reactive oxygen and dismissing the hydrogen atoms. These rose through the atmosphere and, being the lightest atom in the periodic table, escaped the gravitational field of the planet for good. Over one or two billion years Earth would have lost all her water, like her neighbors Venus and Mars, and irreversibly become a dead planet.

At this point, the self-regulating powers of Gaia began to show: life sparked on Earth and changed the course of the planet's evolution. Two types of bacteria appeared that were able to bind Earth's hydrogen: one type was able to gain energy by making hydrogen sulfide near the sea floor.[2] Another type of bacteria invented photosynthesis, thereby taking oxygen from CO_2 and making it available for the creation of new H_2O atoms (water).[3] Earth's water could be saved.[4]

Next, the sea had to be cleaned. The early oceans were a toxic brew, containing high concentrations of metals like iron, chromium, copper, lead and zinc, as well as carbon and other elements. Microorganisms began to use metals as catalysts which speeded up their metabolism. When they died their bodies deposited minute amounts of metals on the sea floor. In the long run, this purged the oceans of the dissolved metals and created sediments with deposits of metal ore. In a similar process, microorganisms created mineral oil and gas reserves (fossil fuels).[5]

Free oxygen was very rare when the Earth was young; it was only a trace element in the atmosphere. Photosynthesizing bacteria, and later their successors, algae and plants, brought up the level of atmospheric oxygen from near zero to the present level of 21 percent. When oxygen finally began to accumulate in the air a whole new chapter of Earth history began. First, this restless element oxidized methane which had previously been dominating the atmosphere. Oxygen processes made nutrients available that had been scarce before and hence triggered the evolution of ever more complex land plants. They in turn created more and more oxygen—most noteworthy are the gigantic forests of the Carboniferous (ca. 360-300 million years ago). When the air had at least 10 percent of oxygen, larger animals could evolve.

Through the huge forests of the Carboniferous, the production of oxygen and extraction of atmospheric carbon virtually shot over the target: levels of up to 35 percent oxygen and a reduction of the greenhouse effect (sinking levels of CO_2 due to its absorption by the megaflora) initiated a massive—and "self-inflicted"—Ice Age (the so-called late Paleozoic icehouse). The first cooling increased the glaciers and polar ice caps, and their increased *albedo* (reflection of solar heat) led to *positive feedback loops* (see Chapter 5) which cooled the Earth further. At the peak of that Ice Age[6] the world was 10 degrees colder. It took Gaia tens of millions of years to get back to a balanced climate and balanced oxygen and carbon levels. That never happened again; all later Ice Ages had astronomical causes (see box).

Irregularities

Some of the astronomical irregularities of Earth's movement around the sun also have effects on Earth's climate:

- Earth's orbit oscillates between nearly circular and slightly elliptical (oval), with a periodicity of about 100,000 years (called *eccentricity*).
- The spin axis of the Earth is tilted against the orbital plane; the angle oscillates between 22.1 and 24.5 degrees on a 41,000-year cycle (obliquity). We are currently at 23.44° and decreasing.

- *Precession* is a slow, circular movement in the top of the Earth's axis, so that its projection points around the zodiac in a 25,770-year cycle; currently the North Pole points towards the North Star (Polaris).

The orbital variations determine the distribution of solar energy around the planet. Eccentricity seems to be the main driver for the periodicity of glacial (Ice Age) and interglacial periods.[7]

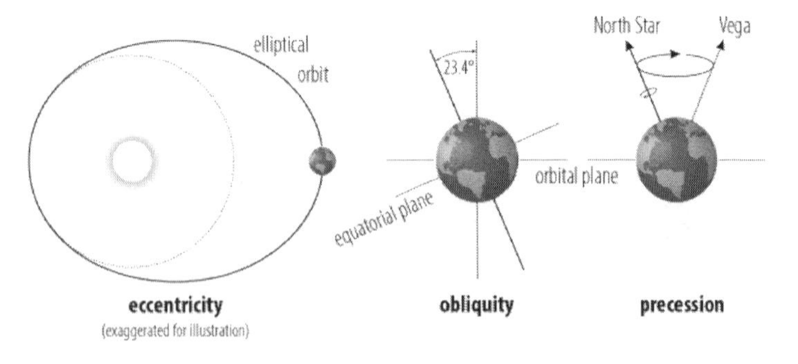

Figure 1. The irregularities of Earth's orbit

Chapter 3

The Elements and Cycles

The elements

Looking at the foundations of our planet in the microcosm helps us to understand the Earth's relation to matter and its role as a self-organizing and evolving entity.

Biotic (biological) and *abiotic* (non-biological) beings alike are composed of complex molecules formed by the chemical elements of the universe. The smallest units of the chemical elements are the atoms. (We can ignore quantum physics here.) It is widely known that atoms are mostly empty space (99.99 percent), and that those "building blocks of matter" consist not of matter but of energy, of positive and negative charge. Following the common atomic model, most atoms by nature do not have the full set of electrons in their outer orbit. And they do everything they can to make it complete. This they cannot do alone, so they interact with each other, thereby forming molecules and ever more complex molecular structures. *So the die is cast for a universe of communication and interconnectedness.*

Aristotle said that the universe is run by two mysterious forces: attraction and repulsion. The entire material universe, all physics, chemistry and biology would not exist without this fundamental dance of opposites. And atoms, like humans, are constantly trying to find fulfilment. We don't necessarily have to think of them as dead, mechanical entities. They have some peculiar characteristics of their own. Of course they always behave in the exact same way when they encounter a particular physical or chemical situation. No matter where in the universe, they follow the laws of physics (so it is believed). But nothing exists in splendid isolation, everything depends on its relationships with everything else. *And atoms are very much about relationships.* Different relationships reveal different sides to them, often astonishing for us and predictable only when you know what's coming.

Take hydrogen and oxygen for example. Hydrogen is the very stuff that makes the sun burn, with a heat of 9,930 Fahrenheit (5,500°C) on the surface and millions of degrees at the core. And oxygen is such a dangerous reactive gas that if we had only 4 percent more of it in the atmosphere the entire surface of the Earth would burst into flame. And what is the outcome when these two wild, primordial fire spirits are combined, two hydrogen to one oxygen atom? Cool, mellow, life-giving water. If that's not a surprise, what is?

There are six elements which are most important for life. They are carbon, hydrogen, nitrogen, oxygen, phosphorus, and sulfur.

Carbon (C) atoms lack no less than four electrons in their outer orbit which makes carbon "a highly cooperative and intensely social being", in the words of ecological scientist Stephan Harding.[1] Complex structures of carbon with oxygen, nitrogen, phosphorus and other atoms form DNA, and proteins, hence *carbon atoms are the chemical basis for life.* On top of that we eat carbohydrates (sugars, starch and cellulose in vegetables) and dress in them (cotton, linen, hemp). Trees are carbon structures, with half of the weight of dried wood being pure carbon. 18.5 percent of the weight of the human body is carbon.

Carbon is present in two of the main *greenhouse gases*: carbon dioxide (CO_2) and methane (CH_4), the third one being water vapor. They affect global climate, plant growth and hence oxygen production.

Hydrogen (H) is the smallest and most simple atom in existence; it only has one electron. A hydrogen atom can either bond with another, *sharing* their electrons as a hydrogen molecule (H_2). Or it can *donate* its only electron for good to a fellow hydrogen, which makes the donor a positively charged *ion* (H^+), and the receiver a negatively charged ion (H^-). "Hydrogen is an airy, flippant creature which would love nothing better than to escape our planet altogether and return to its ancestral domain in outer space" (Harding[2]). Earth would lose all its water if it wasn't for the myriads of living creatures who provide enough oxygen to keep hydrogen from escaping. *Without water there would be no life—and without life there would be no water.*

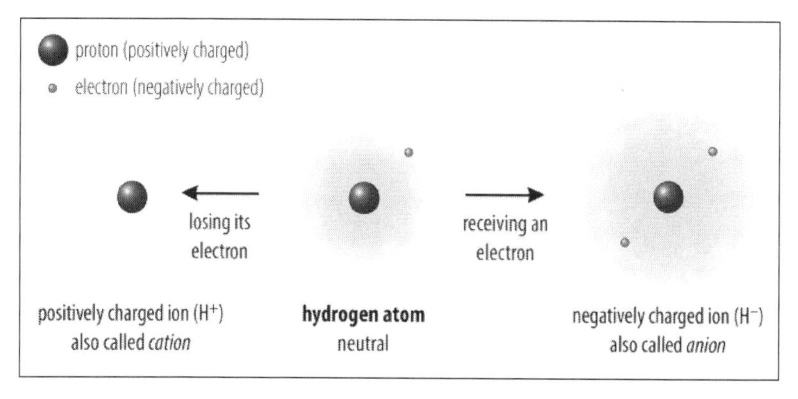

Figure 2. The hydrogen atom and its ions

Nitrogen (N) is star-born and part of the Earth since her very beginning. Chemically it encounters its most stable relationship with itself. When two nitrogen atoms have bonded it's almost for life, and a great deal of energy is necessary to split them again. Nitrogen (N_2) is highly unreactive; if man wants to split it to produce artificial fertilizers, it takes a temperature of 930 Fahrenheit (500°C) and a pressure 1000 times greater than the normal atmospheric pressure. However, soil bacteria perform the same task at ease, at "room temperature", and without contaminating the waters of the living world.

As a key element in proteins and DNA nitrogen is essential for life, particularly for animals: while plants are essentially carbon structures, animals like us are characterized by a nitrogen architecture.

Nitrogen forms 78 percent of the atmosphere of the Earth.

Oxygen (O) has a fierce reactivity with other atoms. Because of this, oxygen was only a trace element in the atmosphere when Earth was young. Over 3.5 billion years ago, cyanobacteria began to produce free oxygen through photosynthesis. They could not offset the oxygen absorption by tectonic and volcanic activity for a long time, but gradually their tireless activity bore fruit and free oxygen levels increased.

Our entire mammalian metabolism depends on oxygen: it is called *oxidative metabolism* because our bodies gain energy from the reaction of our food stuffs with oxygen. But all the while, oxygen is a continuous hazard. As a side effect of this chemistry, our cells are exposed to the toxic effects of *free radicals*. From our distant bacterial ancestors we inherited a number of tactics (antioxidants) to deal with these dangerous substances, but the price is high, *oxidative stress* on the cells is believed to be a key factor of aging. Living on oxygen is like playing with fire...

Among the elements, oxygen is indeed particularly correlated with flammability. Below 15 percent oxygen in the air, nothing would burn (but we wouldn't notice because our brains would shut down). At 25 percent, combustion is instant and so ferocious that even the damp wood and wet leaves of a tropical rainforests would ignite.[3]

Phosphorus (P) is absolutely indispensable for organisms. Not only is it required in the structural framework of DNA and RNA, and for the energy transport in photosynthesis, it also links with carbon and nitrogen to form the adenosine triphosphate molecule (**ATP**) which is the prime "currency" for energy exchange in the body. It gets into circulation in the ecosphere thanks to soil bacteria breaking down phosphate rocks. They were created millions of years ago by marine microorganisms and now they sustain the entire ecosphere.

Sulfur (S) is an essential component of all living cells. In plants and animals, sulfur is present in certain amino acids and in all polypeptides, proteins, and enzymes that incorporate them, and also in vitamins, and antioxidants. Surprisingly, sulfur is either the seventh or eighth most abundant element in the human body by weight (about equal to potassium); a 70kg (150lb) human body contains about 140 grams of sulfur. In Earth's metabolism, sulfur occurs naturally and abundantly in its pure form or as sulfide and sulfate minerals.

Many other elements are crucial for the biosphere. After the six big players described above there are three which occur on a smaller scale but aren't any less important for human and non-human animals:

Calcium (Ca) is vital to the health of the muscles, the circulatory, and the digestive systems. It supports synthesis and function of blood cells, regulates the contraction of muscles, nerve conduction, and the clotting of blood. The Ca^{2+} ion forms stable bonds with many organic compounds, especially proteins. Calcium plays a particular role in the formation and maintenance of teeth and bones.

Although calcium is essential for life it is highly toxic in the free ionic state. Already in the early oceans, bacteria and microscopic algae began to convert the dangerous calcium ions into insoluble calcium carbonate ($CaCO_3$) to make shells for themselves. This did not only reduce the dangerous calcium levels in their cells but also increased protection of their organism immensely. Later, higher animals (including us) adopted the same trick of keeping most calcium out of cell metabolism by using it for bones and teeth.

Iron (Fe) is the central atom in the heme molecule which occurs in hemoglobin (the oxygen carrier in red blood cells), and other heme proteins which participate in transporting gases, building enzymes, and transferring electrons. Iron is also present in certain molecules called cofactors which assist enzymes in biochemical transformations. In Gaia's metabolism, iron is important as a component of various rocks, for example hematite (Fe_2O_3) and magnetite (Fe_3O_4). Surprisingly, living organisms can produce magnetite, and tiny amounts of it can be found in bacteria, insects, birds, reptiles, fishes and mammals who use it for geomagnetism-aided navigation,[4] and it has also been found in various parts of the human brain.[5]

Silicon (Si) tends to make long chains with other silicon atoms and/or bonding with oxygen, for example in the silicate ion (SiO_4). The silicate ions link up again to silica (SiO_2) which occurs in the highly ordered, spiral configurations of quartz crystals. Quartz is the second most abundant mineral in the Earth's crust, being part of the granite bedrock of the continents. The human body needs silicon for the production of elastin (a highly elastic protein for the blood vessels) and collagen (a protein for tissue regeneration), hence silicon is known to be good for the health of nails, hair, bone, and skin.

The cycles

The carbon cycle

When the Earth was young gigantic amounts of carbon entered the atmosphere through the activity of volcanoes and the weathering of volcanic rock. The oceans absorb a large part of this carbon dioxide (CO_2), and rivers deliver more carbon compounds (from the breakdown of rocks) to the sea. The continuous influx of carbon is balanced by marine organisms, especially plankton, algae (like *Emiliania,* see below), and corals, which when they die deposit huge amounts of calcium carbonate ($CaCO_3$, better known as chalk) on the ocean floors. There the bodies become part of the ground layers which over geological time get compressed into sedimentary rock: limestone and chalk rocks (such as the White Cliffs of Dover in Kent) are organic, originating from marine life.

On land, trees and plants filter vast amounts of CO_2 from the air, split the molecules, release the oxygen and use the carbon to create sugar compounds. These can transfer and store the captured solar energy. Plants use them for the maintenance of their bodily functions, and when they die share it with other life forms.

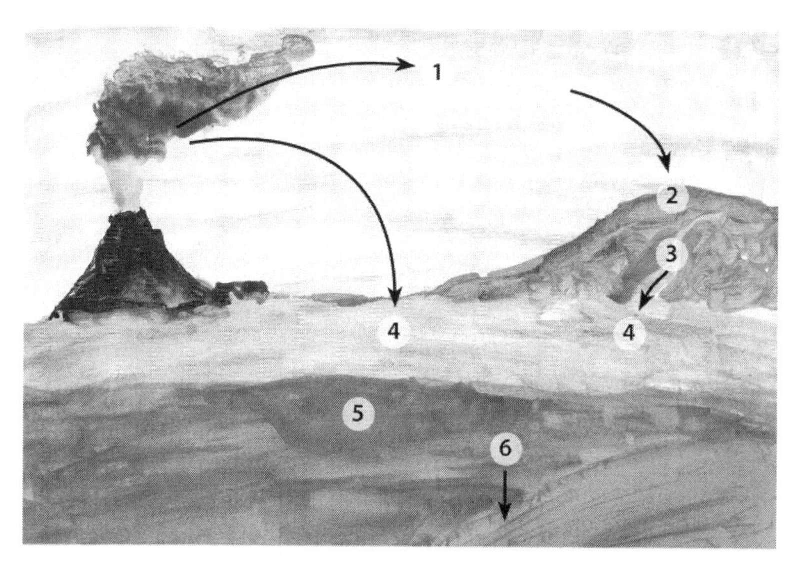

Figure 3. The natural carbon cycle

(1) Volcanic activity releases CO_2 into the atmosphere. (2) Land plants capture CO_2, carbon is bound in their bodies and sequestered in the soil. (3) Rivers carry carbon compounds to the sea. (4) A large part of volcanic CO_2 is absorbed by the sea. (5) Algae and other maritime creatures fix carbon either by photosynthesis or by making shells combining carbon with calcium ($CaCO_3$). (6) Myriads of dead shells sink to the sea floor, and their deposits eventually get compressed into sedimentary rock. (7, not shown) Humans and other omnivores return some carbon to the atmosphere by exhaling CO_2 and emitting methane (CH_4). (8, not shown) With the movement of the tectonic plates, the sedimentary rocks are transported back into the liquid magma of the Earth's mantle where they get recycled. Perhaps after 300 million years a carbon atom might see the light of day again when a volcanic eruption sucks it out of the depth.

Since CO_2 is a primary greenhouse gas (GHG), the biota (forests and land vegetation, plankton, algae, corals, also molluscs and crustaceans), by helping to create carbon sinks, play a huge part in Gaia's management of global temperature. Due to their microflora, the oceans are the world's largest depository of CO_2.

But a planet populated by plants alone could not live long either. In a few million years they would reduce atmospheric CO_2 to dangerously low levels and, by eating the blanket that keeps the Earth warm, would turn the planet to a permanently frozen state (Snowball Earth).[6] That's where we humans and the animals come in, and even more so the "fermenters", the scavengers of the bacterial world who live on the excreted matter or dead bodies of the "primary producer". All of us scavengers use organic matter produced by plants (and other primary producers such as algae and bacteria) and release some of the carbon again into the atmosphere. Thus the carbon cycle finds its equilibrium.

The water cycle

The water cycle gives life to all of us. How marvelous it is that the rainwater finds its way to the rivers which take it to the sea where water evaporates again and forms rain clouds which, by the grace of landward winds, bring us life-giving rain again.

Less known, however, is the role that biomes (communities of plants and animals) play in the water cycle. Water vapor on its own hardly forms clouds, it needs cloud condensation nuclei to kickstart the process. Some dust particles in the air might do the job, but it is particularly algae and forest trees which emanate certain compounds that function as condensation nuclei. Trees release chemicals called terpenes that actively speed up cloud formation. The particles released by pine forests, for example, double the thickness of clouds some 1,000m above the forests. These clouds will carry life-giving rain elsewhere, and on their way the white clouds reflect sunlight (an effect called *albedo*) and cool the Earth.

But trees do not only practice active cloud seeding, they produce the moisture in the first place. Trees pump up so much water from the ground and evaporate it through their leaves that the air above woodlands is saturated with moisture. New clouds can form—not ocean-born but "wood-born"—and drift further inland. In this way, the Amazon rainforest recycles the rain from the coastal belt five to six times until it reaches the Andes on the other side of the continent, some 4,000km away. The Amazon rainforest creates about twice as much water vapor as the neighboring Atlantic Ocean. And while most of the rain from the high seas returns to the big water, the tree-born clouds drift further inland and water an adjoining "belt" of woodland. *Forests are rain-makers. And the clouds they generate are an essential part of the global cooling system.*[7]

In the sea, most algae and also coral reefs emit DMS, a sulfur compound that helps with cloud seeding (see "The sulfur cycle", below).

The nitrogen cycle

Atmospheric nitrogen needs to be "fixed" into a form available to plants. Most fixation is done by free-living or symbiotic so-called *nitrogen-fixing* bacteria who combine it with hydrogen to produce ammonia (NH_3). Two groups of so-called *nitrifying* bacteria successively produce nitrites (NO_2^-) and nitrates (NO_3), thereby guaranteeing a constant supply of N to all plants and animals. Plants can absorb nitrates or ammonium from the soil via their root hairs. Animals get their nitrogen from eating plants. Animal excretions and dying plants and animals return N to the soil where fungi and bacteria convert the organic N into ammonium (NH_4^+). Nitrifying bacteria change ammonium into nitrites, then nitrates again which can be absorbed by plants again. Or it gets into the loving hands of *denitrifying* bacteria who can convert nitrates to N_2, releasing nitrogen back into the atmosphere. Oxidation also returns N from nitrates and ammonia into the atmosphere. (You don't have to *understand* all this, it's meant to give you a *feeling* for how much is going on in a fertile, healthy soil.)

In the sea, the nitrogen cycle is just as important. While the overall cycle is similar, the players are different. For example, the plants' position is held by phytoplankton, and nitrogen fixation is performed predominately by cyanobacteria.

This might all seem insignificantly small to you, but the total nitrogen flux of Earth's metabolism is about 500 megatons a year (for comparison: the volume of the carbon cycle is 800 times bigger).[8]

From Gaia's perspective, nitrogen needs to be kept in the air at its high level of 78 percent. Its collective weight sustains the atmospheric pressure which is a base parameter for all physiological events on Earth. And it serves as a chemical diluent, without which the levels of other gases would rise too much.

The oxygen cycle

Since oxygen is such a highly reactive gas it would mostly be "used up" and is very rare in the atmosphere; as was the case when the Earth was young. But the plant kingdom (forests and algae) and colonies of cyanobacteria in the sea permanently maintain the atmospheric oxygen level at about 21 percent so that it can support animal life. Today, the Amazon rainforest alone produces about 20 percent of the world's oxygen, the algae in the world ocean produce 40 percent. Conversely to the carbon cycle, it is animals (including humans) who use up a lot of the oxygen supplied by the plant kingdom and thus participate in creating a global equilibrium. The energy provided by oxygen allows us to move and birds to fly. Because of oxygen the sky is blue and clear, and the Earth from outer space looks like a beautiful blue marble.

The phosphorus cycle

Phosphorus, the "light-bearer" (from ancient Greek *phōsphóros*), is of central importance to life, being present in the energy-storing molecule ATP as well as in DNA. Its natural source is the weathering of rocks, and its odyssey through the living worlds ends on the sea floor where it becomes part of sediments which eventually get reabsorbed into the magma of the Earth's interior. Hundreds of millions years later phosphorus can reappear on the surface, in newly rising rocks of continental plates. Soil bacteria make the phosphates available for the biota on land where it can circulate for some time. It is rare, however, and its scarcity often limits plant growth.

Those phosphates carried by the waterways to the sea are the only original source of marine phosphorus. Ocean areas far removed from land have limited phosphorus availability, but even in those closer to land phosphorus is a precious element. Phosphorus is the "gold of the biological world."[9] The Phosphorus circulation is rapid. Phytoplankton (from Greek *phyton*, "plant", and *planktos*, "wanderer, drifter") of all sizes, down to the microscopic picoplankton, absorb the biologically available phosphate ion (PO_4^{3-}). After death, phytoplankton who don't get eaten by fish sink into the deep ocean where the precious phosphorus eventually would be lost to the sediment layers on the ocean floor. Therefore a host of bacteria species quickly gets to work to recycle the organic phosphorus from all sorts of dead organic matter, keeping it available to sea life.[10]

The sulfur cycle

The continents continuously lose sulfur because rivers annually carry millions of tons of it to the sea. Marine life has no shortage of this nutrient but Gaia had to develop mechanisms to ensure the return of sufficient sulfur amounts to the landmasses. Most algae and also coral reefs emit DMS (dimethyl sulfide, $(CH_3)_2S$), the most abundant biological sulfur compound emitted to the atmosphere. Its oxidation in the air eventually leads to aerosols which act as cloud condensation nuclei. This extra cloud formation has a global cooling effect and with the rain it brings precious sulfates (SO_4^{2-}) to feed the plants on land. It also has a *multiple positive feedback* on algal growth: clouds locally shade the water (algae prefer cool waters), and they increase wind velocity which stirs the depleted surface waters and mixes them with the deeper nutrient-rich strata. Apart from this immediate reward for the algae (which supposedly led to the evolution of this system) the sulfates on land increase the rate of rock weathering and therefore plant growth—plants which will make more nutrients available which in time will also reach the algae. Thus the sulfur cycle is regarded as mutually beneficial for land- and sea-based ecosystems.[11]

These are not even all of the purely elemental cycles of Gaia. The above introductions are extremely simplified, and the overlaps between them are not covered. As an example for complexity and interconnectedness, here is a brief portrait of a *tiny* creature that participates in the *global* effects of the carbon, water, nitrogen, sulfur, calcium and phosphorus cycles:

Emiliania huxleyi is a single-celled marine alga that lives near the surface of cold oceans among the wider phytoplankton community. Its diameter is 4/1000th of a millimeter (4 microns). Emiliania is the most abundant species of modern coccolithophores ("carriers of little stone berries", how cute is that?), the coccoliths being the tiny wheels which make up their exoskeleton. They're made of calcium carbonate (chalk, $CaCO_3$). Since $CaCO_3$ is transparent, the cell's photosynthetic activity is not compromised by this protective encapsulation.

Even better, its transparency allows coccoliths to scatter more light than they absorb, resulting in surface waters becoming brighter. This means they share the sunlight with the other members of the phytoplankton community. Also, the higher *albedo* of the surface waters reflects more sunlight (a global principle for cooling the planet, see next chapter). Additionally, coccolithophores emit DMS and contribute to cloud seeding. And last not least: When these tiny helpers die they release their nutrients (nitrogen and precious phosphorus) but take the chalk ($CaCO_3$) with them to the seabed. In this way, myriads of coccolithophores over millions of years helped to cool the early planet by sequestering carbon (one cubic centimeter of sedimentary chalk contains about 800 million of them). Emiliania is still doing it.[12]

Chapter 4

Communities and Networks

The matrix of life

The ecosphere which we are a part of is permeated by the Earth's magnetic field "and its occasional perturbations caused by interaction with solar wind plasma."[1] Local geomagnetic fields are also modified by atmospheric electricity (and nowadays by artificial electromagnetic radiation, see "Microwaves" in Chapter 10). Electromagnetic fields influence our fundamental biological processes, even with their most subtle fluctuations. In plants, for example, magnetic fields have been shown to modify seed germination and affect seedling growth and development in a wide range of plants, from grasses, cereals and other (industrial) crops, herbs and medicinal plants, vegetables, fruits, and trees. Science is just beginning to investigate the "definite role of electromagnetic and geomagnetic fields in biological regulation, including regulation of gene expression patterns *in any living being*."[2] (My italics)

All life forms are finely tuned to the Earth's magnetic field which enables them to navigate three-dimensional space as well as geographical distances, to receive signals and communications, and some to even track food sources. Electromagnetic receptors have been found in innumerable species and allow them to respond to even minute changes in magnetic field intensity. There are two main types: a) the magnetite-based magnetoreception system[A] which is used by dolphins,[3] sockeye salmon, algae, migrating birds, ants, bees, and bacteria; and b) the chemical magnetoreception system[B] which is used by some migratory birds, the monarch butterfly, and some other insects.[C,4]

Honeybees, for example, have magnetite crystals located in

their abdomen that serve as a compass to navigate the Earth's magnetic field, and thereby never get lost between their hive and memorized flower stands.[5] Hornets (*Vespa orientalis*), when building their nests, glue a tiny crystal to the roof of each hexagonal cell. These magnetic crystals form a network that helps with the architecture of the nest, and with orientation too.[6] Robins (*Erithacus rubecula*) have a magnetic compass in both eyes and can therefore "see" the Earth's magnetic field and navigate it.[7] Bacteria produce microscopic magnets called magnetosomes[D] which orient the entire bacterium like a compass needle inside the Earth's magnetic field. Loggerhead sea turtles are known for their epic return journeys to the beaches where they were born. They are able to find them by navigating the Earth's magnetic field and seeking out unique magnetic signatures along the coast—a behavior known as geomagnetic imprinting.[8]

Micro-societies

We have seen how much the very atoms of the material world have an inherent drive towards relationship and connection. As creation unfolded into ever more complex structures, vast intricate molecules eventually began to be organized into even richer architectures by a spark we call "life". Bacteria are the early one-celled organisms which ruled the Earth supreme for the first billion years of life (give or take a week or two). And actually, they are still running the show: 13 percent of the biomass of all life on Earth is bacteria.[9] As oceanographer Angelicque White says, "microorganisms largely control the concentrations, distribution, and molecular makeup of nutritional resources in the ocean."[10] And not just the sea, *there is no elemental cycle, no feedback loop, no ecosystem that would work or even exist without bacteria*. They are the foundation of life. Bacteria are the first expression of Gaia.

In the course of evolution, one-celled organisms became more complex and began to tuck away their DNA safely with a protective membrane, thus creating what we call a cell *nucleus*. Species with a nucleus are called eucaryotes, from Greek *eu*, "true", and *karyon*, "nut, kernel". They were able to do more elaborate things with their metabolism, and also connect to form *multi-celled* beings. Bacteria (and their relatives, the archaea) belong to the procaryotes, from Greek *pro*, "before", and are content being a single cell. Inside it, the DNA is floating freely, next to the hustle and bustle of nutrient processing and other activities. They procreate by eating their fill, expanding in size, and then dividing (during cell division, a bacterium also splits and shares its magnetite crystal).

Bacteria performed many evolutionary steps which are still in function today, not just in mammalian bodies (like ours) but also in plants and other life forms. It is bacteria which began to work with sugars, proteins, and the universal energy-carrying ATP molecule. The bacterial cell is protected on the outside by a so-called capsule and a cell wall, lined on the inside with a plasma membrane. *All biological life depends on membranes*, semi-permeable layers of delicate thinness that act as a filter to allow chosen nutrients, proteins and other essential components to enter the cell. Membranes are peppered with tiny holes which let in nutrients but also with other openings to excrete waste products. Each of these "channels" is individually fitted to the respective molecule type, like a key and its lock. Precision is the middle name of cell membranes.

But bacteria can do so much more than just eat and excrete. Modern research has unearthed the most intriguing discoveries about them. They organize themselves into vast, extra-cellular self-making networks (colonies) with such highly sophisticated communication that some bacteriologists have felt tempted to call them superorganisms. "Most bacteria live in communities, often with different cell types carrying out specific metabolic functions, and in order for the whole to work well the multifarious and multitudinous members of the group have to communicate with each other about the complexities of the surroundings which impinge on them, and about the state of the whole community" (Harding).[11]

Their magnetoreceptors enable bacteria to communicate via electromagnetic signals. Many bacteria create thin tubes, so-called nanowires, in their colonies which are used to transfer electric signals via potassium ions.[12] More highly developed bacteria use a "wireless" version of intercellular communication. By moving electrons around DNA loops, bacteria such as E. coli can create radio frequencies (0.5-1.5kHz) and send wireless broadcasts..[13]

In addition to electromagnetic signaling, bacteria communicate with special molecules, similar to insects and plants who use pheromones for far-distant communication. Bacteria have molecular "words" for the affairs within their own species, and they have other signals which are understood by all bacteria, like an international trade language among humans. Thus every single bacterium knows how many of its own species and how many other tribes are out there, and what they are doing. Their type of communicating with each other is called *quorum sensing*. Example: the rod-shaped soil bacterium *Myxococcus xanthus* lives on decaying plant material. When the colony has a feast day, the single members spread out along tiny slimy trails, all the while feasting, growing, dividing. But when the food source comes to an end, they know via quorum sensing that it's time to regroup. Every individual backtracks along its trail and they all end up in one big mound where most of them commit suicide. The dying cells release their nutrients which gives the few survivors the energy to produce resistant spores before they die too. The spores are able to wait until favorable conditions return.[14] What does this actually involve? An individual follows the group call to form the mound, then has to assess its own interior and tell the community about its state. Only then the group is able to decide which ones will donate and which ones will receive the remaining nutrients.

The startling complexity of bacterial communication has led experts to discuss "bacterial syntax", "social intelligence", "common knowledge" and "collective memory", "group identity", "learning from experience", "improving themselves", and "group decision making."[15] It comes as no surprise, then, that bacteria keep their DNA as a kind of *open source* program (to use computer language). After decades of heavy attacks by man-made antibiotics, those bacteria who survived onslaughts have finally been able to write the immunity mechanism into their DNA and share it with others—first in their tribe, then cross-species! Since then, health professionals worry about multi-resistant "super bugs".

Teaming up

One sunny September day (just joking) about 2.5 billion years ago, a bacterium found itself inside another cell (either another, bigger bacterium or an early eucaryotic cell, we will never know). But instead of following their food chain habit (we don't know either which one of them was the predator) it happened there and then—in a split second before the lightning-fast chain reactions of molecular digestion kicked in—that they co-opted to try to live together in symbiosis. The host would share its bountiful cytoplasm full of nutrients, and the "guest" was particularly good at creating energy in the form of ATP, thus boosting its host's levels of energy and joy. If more energy was needed, the small visitor was even allowed to procreate and have akin company in his host cell. Today, we call them **mitochondria**.

As cells do, they tell each other. The idea was taken up and eventually became a success across the biosphere. Later, when complex organisms developed, the incorporated mitochondria were always a part of every cell. Still today, every fungi, plant and animal cell has mitochondria partners inside it. Every human cell has quite a few of them; active tissues like brain or heart might have about 300 mitochondria per cell, liver cells have up to 2,000. *And mitochondrial DNA is still entirely separate and independent from human, animal or plant DNA.* We are like spaceships with countless little guests who give us the energy to function in the first place.

Later in evolution, when the first plants got the idea of what they wanted to do in life (to photosynthesize), instead of reinventing the wheel they allowed photosynthesizing cyanobacteria to come inside their cells for a similar symbiosis. They became the green **chloroplasts** which still do all the work in green leaves today, turning solar power into sugars. Sweet, isn't it?

Networking

Fungi are best known for their fleshy fruiting bodies which can be very delicious and nutritious to eat (or very poisonous). But fungi are much more. *Fungi create the most important nutrient transport systems on the planet.* Without them there would be no forests or savannahs, not even grasslands. The Earth's land surfaces would consist only of exposed rock in wet places covered with expansive layers of bacterial colonies.

Fungi create long tubes, so-called hyphae, which transport nutrients through the cellular fluids. Hyphae are microscopically thin; their width varies between 1/500th to 1/100th of a millimeter. They interlink to form a vast network called mycelium. These networks can cover acres and acres. The largest known mycelium in the world covers an area of 2,400 acres of forest in Oregon, and is estimated to be 2,200 years old. Mycelia grow extremely dense, a single gram of soil can contain over 3,000 feet (a kilometer's) length of microscopic hyphae. In grasslands, the biomass of underground mycelia matches or even outweighs that of animals.[16]

An important evolutionary step for fungi was to team up

with higher plants to form a symbiosis: a close relationship that is mutually beneficial. Fungi provide plants with large amounts of water and with soil nutrients (phosphorus, nitrogen, copper, zinc, and others) which would otherwise not be accessible to plants, either because they're out of reach or chemically unavailable. In return they receive sugars which they cannot produce themselves as they don't have the capacity to photosynthesize. For this symbiosis, called *mycorrhiza,* fungi intimately bond with the plant roots. "Fungi specialize in interconnecting other living entities" (Harding),[17] and we know now that they can share nutrients or sugars from one tree to other trees hundreds of yards away. They manage the nutrient distribution in their ecosystem effectively with nutrients able to flow in both directions of the hyphae.

And so does information! Mycelia have been shown to "learn" from experience and "memorize" their adequate responses. For long, fungi had been viewed as mechanical, predictable systems, but they reveal inherent intelligence and creativity. When a mycelium explores its surroundings (by growing new hyphae) it clocks patches of edibles, and also works around obstacles. Hyphae bundle up to create superhighways connecting to reach food fields. Laboratory tests have shown how a slime mold can find the shortest way through a labyrinth.[18] In a BBC documentary slime molds even recreate the efficiency routes of the Tokyo underground system, or motorway networks in the UK and the USA.[19] No wonder then that mycologists speak of "brains in the soil" (Alan Rayner).[20] Fungi are over 1,000 million (= 1 billion) years old. Like bacteria they originated in the sea, but they arrived on the rocky land masses about 500 million years before humans appeared on Earth.

Another type of fungi does not work underground but penetrates the wood structures of living trees. These and the mycorrhizal fungi represent an essential part of the trees' **immune system** because they shield their host plants from bacteria and other fungi which wouldn't be so benevolent. Yet another type of fungi specializes in breaking down dead wood into its components. When the vast forests of the Carboniferous changed the atmosphere of the Earth for good, this type of fungi hadn't fully evolved yet. Gigantic amounts of wood sunk undigested into ever deeper layers of the soil and eventually turned into what we know today as coal. Since then, wood-decaying fungi, like their mycorrhizal siblings, play a crucial role in the carbon cycle and hence in global temperature regulation.

There are more species of fungi on Earth than of plants and animals together.

Another surprising leap for life occurred when fungi and algae combined to create lichens. **Lichens** are those humble orange, grey or black patches on rocks and trees that are either flat and crustose, or foliose, looking like lettuce leaves, or fruticose, looking like beards or strands of hair. All of them have photosynthetic algae inside a strong fungal body structure. They pioneer barren rocks and slowly fertilize the ground, first for mosses to follow, then for more complex plants and plant communities. Silicate rocks weather ten to one hundred times faster with lichens working on them than without. By extracting carbon from the air and combining it with calcium from the rocks, they help to cool the Earth.[21]

Waste management

In nature, the waste products of one species are food for another. *There is no "waste" in nature, endless recycling is the magic of Gaia's sustainability management.* Waste products even "create" new species because they provide a type of food that wasn't there before, and trigger colonies of waste users to adapt to the new habitat and re-shape themselves over an evolutionary time-scale.

Footnotes

A. Magnetite-based magnetoreception uses ferromagnetic crystals of magnetite (Fe_3O_4) which orient themselves in geomagnetic fields. Exposed to local changes in polarity, their position and distribution changes, which can cause a switch in the ion channels on cellular membranes.

B. Chemical magnetoreception is a light-dependent magnetic sense which is mediated by an ultraviolet/blue light photoreceptor flavoprotein called cryptochrome.

C. A third system works with electromagnetic induction and can only be found in sharks, stingrays, some fish, and other aquatic species. Their electroreceptive organs are capable not only of detecting electric fields of other creatures but also of sensing magnetic fields.

D. Magnetosomes are membrane-enclosed magnetite crystals. The assembly of highly ordered magnetosome chains is under genetic control and involves several specific proteins. Magnetosome architecture along a cytoskeletal structure is one of the most complex that has ever been found in bacterial cells.[22]

Chapter 5

Feedback Systems

As an example of self-regulating feedback loops let's visit once more our old friends, *Emiliania huxleyi* and the coccolithophores. They help in regulating the temperature of the oceans. But they cannot overdo it or underdo it because of simple feedbacks: they thrive in warm waters and release DMS (see "The sulfur cycle", Chapter 3) into the air whose nuclei help to produce thicker clouds to block the sun. This cools the ocean, the algal population decreases and in turn the amount of clouds decreases as well. With fewer clouds blocking the sun, the temperature rises again, and algae density increases again.

Systems science distinguishes between positive and negative feedbacks. Positive feedbacks have an enhancing effect on the next link in the chain reaction, negative feedbacks decrease the effect, they put a brake on and calm things down. A metabolic cycle needs *at least one negative* feedback or it spirals out of control: sunshine warms water > sunshine and warm water boost algae population > more algae seed more clouds >> *clouds block sun*.

In this example, if clouds did not block the sun then algal bloom would go completely out of bounds. The way it is, there is a negative feedback in the circle, creating a balanced pendulum motion.

Another dynamic: from each pair of carbon atoms taken up by the coccolithophore cell for the production of its plates one is used to produce the chalk ($CaCO_3$) and the other is released into the water again. Since carbon in water turns into carbonic acid, and acidification is a global heating effect, some researchers around the turn of the millennium wondered if coccolithophores contributed to global heating.[1] But overall, the opposite is true:

before long the other carbon atom via its chalk journey will be permanently sequestered on the ocean floor; coccolithophores provide a sink for emitted carbon, mediating the effects of greenhouse gas emissions.

A third dynamic: During the complex plate production, coccolithophores constantly pump H^+ ions out of the cell. H^+ ions turn their environment more acidic, and when exterior acidity goes too high, the algae sense that and halt production. This is a perfect negative feedback, a safety valve against an imbalanced algal bloom. However, today the oceans rapidly acidify due to man-made carbon emissions, and the resulting levels of carbonic acid begin to impair the health and safety of our little friends whose channels for hydrogen ion excretion are rendered dysfunctional. We know that there is a point for each algae species or population where exterior acidity levels get so high that the living beings will shut down and disappear.[2] This would represent a **tipping point**. (More about tipping points in Chapter 14.)

An element in many climate feedback loops is the so-called **albedo** (from Latin for "whiteness"). This term stands for the measure of reflection of solar radiation back into space. Since solar heat reaches us in the form of radiation it can be reflected like light. As we know from daily life, the more light-colored a surface the more does its reflection blind our eyes. Dark surfaces absorb light and heat and warm up much quicker.

That's why the cloud seeding of forests and algae populations is so important to global climate. Atmospheric water vapor traps heat but white clouds reflect it and keep the Earth cool. The polar ice sheets are of paramount importance, for their albedo, and for other reasons as well. When polar sea ice melts the Earth doesn't only lose out on the reflection from the shining white ice, but the newly exposed water surface is dark and absorbs most of the solar heat. White surfaces like snow, ice and clouds have a high albedo, reflecting 80-90 percent of solar energy, water has a low albedo and reflects only 5-10 percent.

But a low albedo can also be an advantage for an ecosystem. Warming the ocean water is also important because a higher temperature *gradient* (i.e. the more difference in temperature) between top and lower strata the better will these layers mix, thereby exchanging their nutrients for sea life. And in the boreal forests of the Northern hemisphere (North America, Scandinavia, and Russia) the evergreen dark conifers have a particularly low albedo which helps them to warm up quicker at the return of the sunlight, thus bringing forward the coming of spring to these regions where the summers are short.

After so much about bacteria, microscopic algae, and bugs, let's move up all the way to the top of the food chain and look at some "proper" animals: visible, furry, with eyes we can look into.

Apex predators like wolves take the elderly and sick among their target species, thereby saving them some considerable suffering. Also, by keeping the genetic pool of their prey population in top condition, they increase the survival rate of their "victims". Without the hunters, herbivore populations increase, sometimes to dangerous levels running the risk of infectious diseases which one day could wipe them out for good. Another imbalance of unhunted herbivores is over-grazing. Most foresters around the globe have grave worries about the natural rejuvenation of the (temperate) forests in their care because the seedlings are seriously beleaguered by browsing deer. This is different in forests guarded by wolves, lynx and brown bears. They are *keystone species* for their ecosystems.

The early ecologist Aldo Leopold was the first to write about the wider ecological feedbacks of top predators. In 1949 he coined the term "Thinking like a mountain" and described how the disappearance of wolves would lead deer to overpopulate and over-browse a mountainside until it is barren. The fertile soil, not protected by its plant cover anymore, would erode and be lost to the once abundant slopes. This was a fate that neither the wolf nor the mountain had deserved.[3] Eastern European foresters had known this all along, an old saying goes "Where the wolf walks the forest grows."

Over half a century later, Leopold was proven right for good. After an absence of nearly 70 years, the grey wolf, *Canis lupus*, was reintroduced in January 1995 into several areas in the northern Rocky Mountains of the United States, among them the Yellowstone National Park. The presence of wolves substantially changed browsing behaviors of their prey species, in this case predominantly wapiti (in the USA commonly known as elk), who soon stopped chomping their way through the valleys and gorges where wolves could easily ambush them. Native flora was able to re-establish and re-grow, thereby increasing biodiversity by providing food and shelter to a growing number of plants and animals. With the wolf back, vegetation recovered also along the riverbanks, thereby decreasing their erosion. The stabilized rivers meandered less, the channels deepened and small ponds formed. Woods of aspen, willow, and cottonwood established themselves in only a few years, and in turn came the birds. Because the wolves decimated the coyotes, rabbits and mice increased which brought more hawks, weasels, foxes, and badgers. Ravens and even bald eagles took care of the carcasses left by wolves. Down by the water beaver numbers increased, and their architectural work in turn created habitats for other species, such as otters, ducks, reptiles, amphibia and fish.

This top-down cascading effect instigated by a predator is called a *trophic cascade*, and it can, as in Yellowstone, even alter the geography of the landscape itself. Wolves are an integral part of the ecosystem, even small populations create niches that other animal and plant species can inhabit.[4]

The UK, sadly, is far too densely populated for considering

the reintroduction of the wolf as of yet. The Germans thought the same of their densely populated country, but the decision was made for them by the wolf. In 2001, the first wild pair bred again in the country, after the species had been absent for 150 years. In less than two decades, due to further influx from Russia and the Carpathian Mountains, and to the newborn native wolves the German wolf population has grown to about 300 animals who live in 60 packs.[5] Apart from farmers and the hunting lobby, the human population on the whole is sympathetic to the return of this iconic species.

Another predator and keystone species is North America's smallest marine mammal, the sea otter. Hundreds of thousands of sea otters once inhabited the North Pacific coasts but were annihilated by the fur trade in the 18th and 19th centuries. Sea otters only survived in a few isolated populations, but after the Fur Seal Treaty was signed in 1911, numbers could slowly increase. Conservation efforts helped by releasing young otters into the wild.[6] In the Elkhorn Slough south of San Francisco, struggling seagrass beds in the estuary recovered and increased more than sixfold in just three decades.[7] How can otters possibly affect seagrass health?

Sea otters feed on a variety of marine invertebrates, including sea urchins, clams, and crabs. With fewer crabs, the California sea hares, a sea slug (*Aplysia californica*), increase in number. And the slugs in turn feed on the algae growing on and suffocating the seagrass. Healthy and clean seagrass ecosystems provide nursery habitat for young fish and invertebrates, and they even counteract erosion by holding down and trapping sediment. And by absorbing atmospheric carbon and burying it in its roots, they act as a carbon sink.

Sea otters also feast on striped shore crabs (*Pachygrapsus crassipes*) which burrow into muddy banks and feed on the marsh roots, thereby accelerating the erosion of the marsh. By eating the crabs, otters again help to slow down erosion. In the

Aleutian Islands in Alaska, sea otters keep the amounts of sea urchins at bay. Urchins feed on kelp (a special type of marine algae), and without population control by predators, can even erase it entirely. In contrast, with sea otters on patrol kelp forests can thrive. Kelp forests are recognized as one of the most productive and dynamic ecosystems on Earth.

Chapter 6

Diversity, Complexity, and Abundance

Ecosystems are the more stable the more species they contain. Biodiversity is not just luxury or a whim of nature, it is a safety measure that ensures life's continuity. Life is abundant—and has to be! There is no such thing as partial occupation of a planet; if life is to sustain itself on its planetary home it has to be strong in numbers and diversity. Once life becomes abundant enough to have considerable environmental effects, it takes over its planet home: "life in the universe, in general, is likely to be a planetary phenomenon," says Prof. Eileen Crist.[1] Lovelock has always insisted that "organisms are not mere passengers on the planet"[2]—"they are more like pilots," finishes Crist.[3]

Life and its "environment" are closely coupled. And **evolution** concerns Gaia as a whole not the organisms or the environment separately. Life itself helps to shape its environment, and evolution favors species that influence their surroundings in a way that helps their offspring to flourish. *Darwinian evolution is about the long-term development of species and Gaia,* not about individuals fighting each other. Evolution is about how species develop in and with their environment over time spans of tens of thousands of years.

As much as a population enhances its chances of long-term survival by creating environmental conditions which are favorable to its offspring, so the opposite is true. If a population degrades its surroundings its future prospects will be dim. "Gaia theory proposes that organisms inflicting damage on their surroundings will eventually reap harsh consequences when feedback comes back to haunt them." (Crist)[4]

For physicists, the universe will eventually "run out of steam" (following the second law of thermodynamics, entropy).[A] But life turns the cards and makes existence unpredictable. *Life is the antidote to the mortality of the universe.* Life continuously reverses the increasing entropy of all existence by organizing ever more complex organisms and superorganisms, gathering experience and storing information. Metaphorically, says Lovelock, the "most amazing property and characteristic of life is its ability to move upstream against the flow of time. Life is the paradoxical contradiction to the second law of thermodynamics [...] Even more remarkable, this unstable, this apparently illegal, state of life has persisted on the Earth for a sizable fraction of the age of the universe itself."[5] For well over three billion years, Gaia has been increasing her degree of order and complexity. Earth truly is the planet that breaks all the rules.

Depending on our perspective, the tendency towards ever more complex structures is not just a purely biological phenomenon but already inherent in the nature of the atoms. So we could say that the universe right from the start has an inbuilt direction to more complexity, higher degrees of order, eventually sentience, and, as evolution shows, also self-awareness. Perhaps that is why the human being is so important to Gaia.

And there she is, our blue planet. For 3.8 billion years she has kept the temperature, the ph levels, the salinity and all the other parameters favorable for life. Gaia is strong. Life is so strong on Earth that neither the comet that extinguished the dinosaurs nor a possible disaster unleashed by human ignorance could *fully* exterminate it. But we can threaten the survival of our own species (which is what we are currently doing with full steam) and we can harm Gaia badly and throw her evolution back by millions of years.

Life is unspeakably precious. *And for us Life has to be synonymous with Earth,* for we don't know of any other. Even if we found another living planet with one of the new super-telescopes we could not get there. That would be missing the point entirely anyway because *we are a part of Earth*.

Prof. Eileen Crist summarizes the essence of the ecosphere as *diversity, complexity, and abundance*, qualities that "have been captured peerlessly by the perspectives of Darwinian evolution, ecological science and Gaia enquiry." Together they "hold the potential to create a Zeitgeist of deep understanding and harmonious living on Earth. The tendency of life to become increasingly complex, and increasingly abundant has, over the course of eons, created and recreated a living Earth that... can be celebrated as a *cosmos*[B]—a world of intrinsic order and beauty."[6]

Footnotes

A. The second law of thermodynamics says that when one form of energy is converted into another, a certain proportion always escapes as heat. In other words, all physical processes logically move towards thermodynamic equilibrium, until no more energy can be gained anywhere. This steady state is called the state of maximum *entropy* (i.e. the degree of order is zero). Everything in the universe is increasing in entropy, i.e. it increases in disorder. Living organisms require food as a constant input of energy to offset this physical law and maintain their high inner levels of organization (i.e. their low entropy).

B. Unlike the term *universe* the word *cosmos* implies viewing it as a complex and orderly system or entity. It is the opposite of chaos.

Part II

Global Disruption

Symptoms, Causes, and Reasons for Global Disintegration

Now, on every continent and in every sea, climate disruption is becoming the new normal. Human conduct is also leading to severe biodiversity loss, changing animal-human interaction and distorting ecosystem processes that regulate our planetary health and control many services that humans depend on. Science is screaming to us that we are close to running out of time—approaching a point of no return for human health, which depends on planetary health.
– António Guterres, UN Secretary General, 2020[1]

Nobody doubts anymore that an essential ecological crisis is upon us. The symptoms of planetary dis-ease and the constant bad news overburden most of us, and the wide-spread collective reaction to bury our heads in the sand still lingers on. There is enough pleasant distraction around, after all, why not enjoy life as long as we can?

However, living with unattended worry and suppressed fears isn't exactly enjoyable. And living with our head in the sand gets quite dark after a while—where is the light of day? We all have got used to living in dim twilight, in a daily grind of truncated passions and suppressed empathy, oscillating between contentment with what "little we got" and anxiety and deep fears of losing it. A *happy life* is a different thing. We would love to find meaningful opportunities to make the world a better place but we feel helpless as "reason" keeps telling us that we are too insignificant to make a difference. So, are we truly meant to just shut up, sit tight and wear sackcloth and ashes while the Earth is being torn to pieces?

There is another way. Of multitude, of color, of empathy, of love. But it starts with looking at *what is*, and making sense of the situation. I am afraid it is not going to be nice; we will see why we looked the other way for so long. But only by facing our fears can we overcome them. Only by talking about a problem can we find solutions together, and take responsibility into our own hands.

So here's an overview of what's going on and how the different causes and effects are interlinked. And suggestions for steps we can take.

Note: The intention of listing the damage is not to depress anyone but to invoke the fire of our just wrath and of our love. We need all our true emotions to fuel our passion for life!

Planetary boundaries

Human populations too can be part of the intrinsic beauty

and harmony of nature. They can even act as a stimulus for biodiversity, "changing the environment" is not necessarily a bad thing: beavers build dams and flood wetland areas, thereby creating new habitats in the interplay zones of water and land which allow a multitude of species to flourish. Forests are the most species-rich ecosystems on land, and human settlements along their fringes can create new types of local ecosystems which allow for even more biodiversity. Gardens and small organic farms offer a stable habitat for a wide spectrum of species which might otherwise only occur in the short time spaces of forest clearings. Among plants, elder (*Sambucus*), rowan (*Sorbus*), plantain (*Plantago*) and dandelion are among those which find much better living conditions in the cultural landscape than in wild nature. Among animals, field hare, barn owl, and northern lapwing (*Vanellus*) benefit from human presence. Species that seek and thrive in cultural landscapes are called hemerophiles.[A]

However, species such as domestic rats, cockroaches, headlice, and many pathogens also benefit from human settlements. In (too) densely populated localities (and coupled with other factors such as bad nutrition), pathogens can explode into epidemics or pandemics which represent a natural boundary to overpopulation aka dominance of one species that threatens biodiversity (see "Pandemics", Chapter 9). A fundamental law of nature is balance.

And we have an old habit of violating this balance. Sometime between 10,000 and 12,000 years ago, with the dawn of agriculture, homo sapiens began to claim more than its rightful place in the interconnected web of life on this planet. Throughout the historic timeline of evolving "civilization", the disconnection from nature, along with greed and hubris, continued and magnified. The different stages of homo sapiens losing its mental balance include the deliberate misinterpretation and abuse of religion ("You shall take dominion of the Earth"), the era of "enlightenment" which furthered mental disconnection from nature and abandoned empathy with living things, and industrialization. The current stage of global capitalism and its extractive economy is the pinnacle of what is possible on the collective road to self-destruction. Digitization, aka remote digital control of homes, farms, wild nature, and ultimately human consciousness, may dominate the short, grand finale.[2]

What's lacking is the realization and acceptance of planetary boundaries. Any human society aims to raise its industriousness above the poverty line, and rightfully so. Trade and production (of food and useful goods) are means for material prosperity, and a solid material base enables us to unfold the higher potential of the human being: to develop a flourishing social, cultural and spiritual life, and insights and empathy that can also benefit the more-than-human life such as plants and animals, even entire ecosystems. The human species could be a blessing for this Earth.

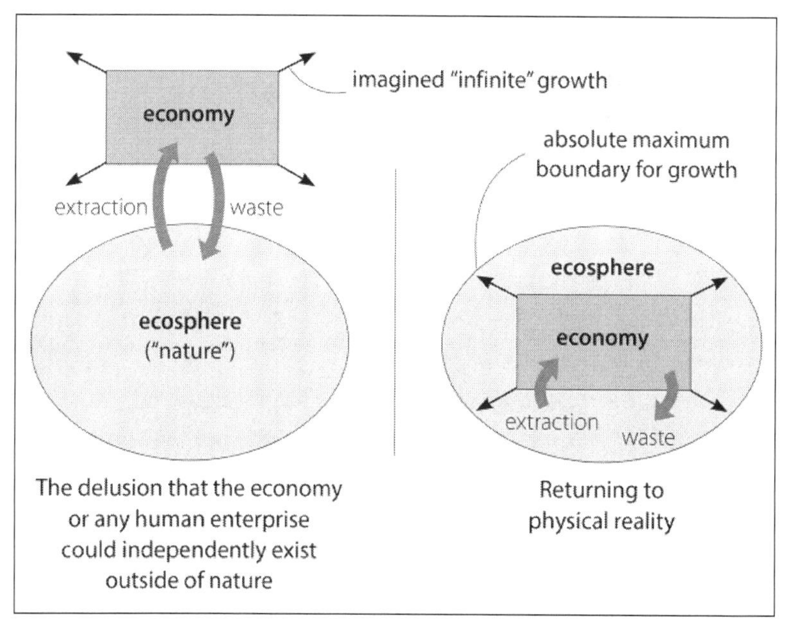

Figure 4. The infinite growth model: delusion and reality

But instead, fear and greed got the better of us. We have long ignored that any given landscape has only a limited *carrying capacity*. No ecosystem can sustain an exploding population of any single species indefinitely.

For too long, the economic model in capitalism was the growth economy which demands never-ending growth of GDP (gross domestic product). This is a collective delusion that sees the human being decoupled from nature. "Resource extraction" and waste dumping are expected to continue ad infinitum. But economy cannot expand forever. **Infinite growth is not possible on a finite planet.** Economy can never exceed the boundaries of ecology.

> As a "choreographed hallucination", the neoliberal paradigm contributes significantly to planetary unraveling. Neoliberal thinking treats the economy and the ecosphere as separate systems and essentially ignores the latter.
> – Prof. William E. Rees, University of British Columbia, 2020[3]

In 2012, the economist Kate Raworth of Oxford University's Environmental Change Institute began to develop the "Doughnut model" as a new way of approaching economics in the 21st century.[4] She says that the aim of economic activity should be "meeting the needs of all within the means of the planet". Instead of economies that need to grow, whether or not they make us thrive, we need economies that "make us thrive, whether or not they grow".[5]

The Doughnut model is a visual framework for "sustainable development" which acknowledges the reality of planetary boundaries as well as social boundaries. The safe and just space for humanity to operate in is depicted like a doughnut or lifebelt. The boundary on the inside of the circle is the social foundation with its twelve categories (food, water, housing, healthcare, etc.). A shortfall of economic activity produces a shortfall in some or all of these areas, resulting in poverty and suffering. The outer border comprises of nine ecological ceilings (climate, freshwater, soil, air, etc.). **Overshoot** results in the destruction of habitats, ecosystems, livelihoods, climate stability, and much more.

The current form of economy fails miserably in both directions. Hunger, poverty, lack of freshwater, among other essential basics, are suffered by billions of humans. And the massive overshoot that violates the planetary boundaries has ushered in the geological epoch of the *Anthropocene* ("age of human")—with mass extinction, climate breakdown, and an

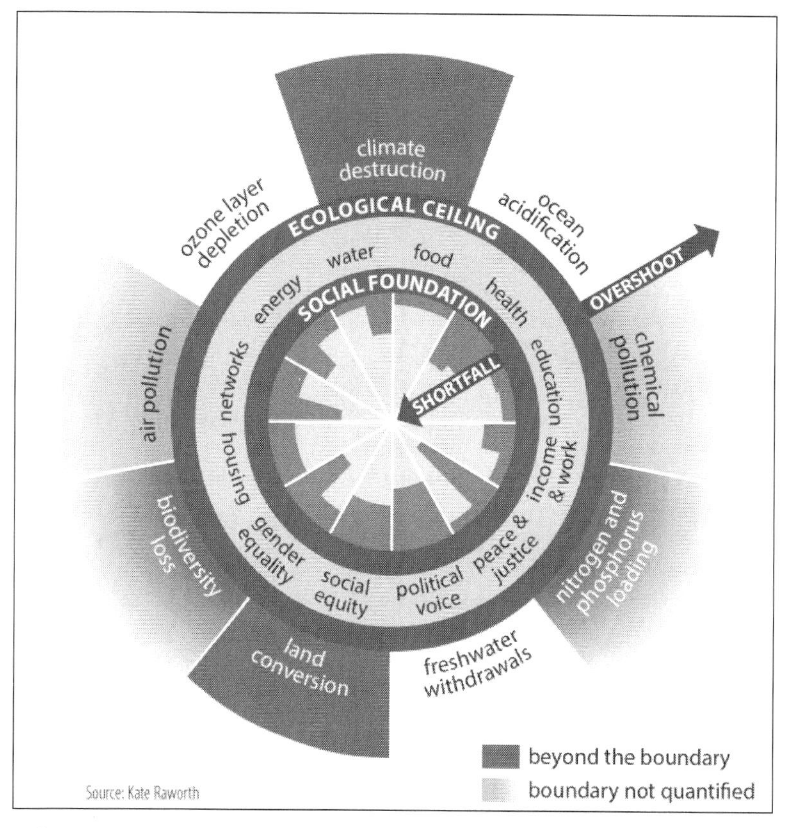

Figure 5. The Doughnut model of social and planetary boundaries after Raworth 2017

omnipresent pollution of heavy metals, radioactive as well as plastic particles which all have begun to leave their mark even in geological strata.[6]

The doughnut model is accepted by the UN as a compass to further their 17 Sustainable Development Goals (SDGs). (See box Chapter 14.)

Reclaiming our language

Before we enter the ecological arena and diagnose the crisis, before we can raise our voice for a healthy planet, we'd better practice watching our vocabulary. We live in a time when battles for ecosystems are predominantly fought with words. Words influence the perception and consciousness of a thing. Words lead to permissions, decrees, court rulings. Misrepresentations and euphemisms abound in the mass media. The industry and economy lobbies have always dismissed conservationists as either naïve dreamers or hysterics. However, this reflex cliché is fading in the age of information leaks and whistleblowers. But it is not surprising that the most important terms are still being twisted in our mouths. The hijacking of key words has been going on steadily for decades. Here are the most important cases and some interesting eye-openers.

"Environment"

When we say we want to do something for "the environment", we have already lost! Because this term keeps cementing the fundamental belief of our current form of civilization, the belief in a separation of Man and nature, and that Man is center-stage, and that everything "else" revolves around him and has to serve him (anthropocentrism). Man is the alpha and omega, the first and the last. The term *environment* (from French *environ*, "to surround, to envelop") describes something like an accessory, and as such it can be exchanged, manipulated, exploited and destroyed. Environmental protection degenerates to mere management and bookkeeping of this destruction, and often focusing on the protection of only that which can serve Man better if it's not totally broken. The terms **biosphere** and **living world** are a much healthier approach. And since the biotic (biological) and abiotic (non-biological) realms work hand in hand, the most correct and inclusive term is **ecosphere**.

"Resources"

Long ago we spoke of *natural treasures*. That is beautiful because it gives appreciation to the wonders of the Earth and at the same time it leaves completely open whether one wants or has to touch them at all. Yes, the Earth has many treasures, and unfortunately we have to use a few of them—very tenderly—for our well-being. The term "resources", on the other hand, has an almost demonic quality. It seems to come directly from the language of neoliberal banking, because it means nothing else than the *market value* of these natural treasures. And you *have* to exploit them, that's inherent in this term. I shudder when I hear this word. Let's talk about **natural treasures** again.

"Climate change" and "global warming"

It sounds so poetic: climate change! Hasn't the Earth always been changing? Isn't that part of nature's harmony? Isn't Greenland called Greenland because it used to be green? I give the oil industry 10 out of 10 for this creative wording, which lulls us all into a feeling of restful peace, time and again.

However, we should mercilessly delete this term from our vocabulary and call a spade a spade. For what humanity is doing, a more honest term has long since established itself among global experts: **climate disruption**. Because what happens is not a "change", nor a "crisis" (crises are temporary and often solve themselves, somehow), but massive, destructive interventions in planetary life structures and feedback systems. It is a pure orgy of demolition, which can also be called *climate* *destruction*. Since 2019, the UN speaks increasingly of **climate crisis** and **climate emergency** rather than climate change. Similarly, **global heating** increasingly replaces global warming.

"Sustainable"

It was a huge gain for international conservationism when the term *sustainable* emerged. Now there was an objective term that could only be used for things that guaranteed the long-term well-being of the Earth and all her creatures. But very quickly the economy lobby began to use it for anything that is "worth the investment". Even a bucket-wheel excavator that amortizes swiftly is called "sustainable", or an oil share that generates a permanent profit. Additional confusion is caused by instances such as those praising single-use chopsticks made of wood as "sustainable" because they are not made of plastic (although for them whole forests are cleared), and praising plastic chopsticks as "sustainable" because they are not made of wood. When indeed the only truly "sustainable" way would be to bid farewell to single-use products and consumerist throwaway culture.

Furthermore, the terms *sustainable* and *non-sustainable* are being used deliberately to avoid the most important subject of our time: **ecocide**. Whenever you want to bring across clearly what you mean, don't say, "This is non-sustainable," instead say, "This results in ecocide."

"Ecocide"

Ecocide is a criminal human activity that violates or destroys ecosystems or harms the health and well-being of a species (including humans).[B] The word was first recorded following the massive damage and destruction of ecosystems in Vietnam perpetrated by the US Army with the widespread use of the plant defoliant "Agent Orange".

In the definition submitted to the UN Law Commission in 2010 by the Earth lawyer Polly Higgins, ecocide is "the extensive damage to, destruction of or loss of ecosystems of a given territory, whether by human agency or by any other causes, to such an extent that peaceful enjoyment by the inhabitants of that territory has been severely diminished."[7]

Eco means house, abode, dwelling (from Greek *oikos*) and refers to our home planet; *-cide* (from Latin) means killing, destruction. So ecocide denotes the serious harm or destruction of an ecosystem or of any species therein. Ecocide is the missing crime in the Rome Statute of the International Criminal Court; its inclusion has the power to significantly change humanity's current course of self-destruction.[C]

"Development"

A word which is nowhere officially defined but, exactly because of its deliberate ambiguity, extensively used in international affairs to mask economic and strategic interests as humanitarian and/or ecological activities. (See box at the end of Chapter 14.)

"Predator"

For most people, a beast or bird of prey *steals* something. Don't all carnivores steal the life of their prey, and run off like thieves in the night? Well, don't all herbivores destroy living plants? Outdated names like this go back centuries, when early science was steeped in Biblical morals. And we have a blindspot for ourselves. Sharks are often called the biggest predators in the world—annually they kill about 17 humans. But every year the industrial fishing fleets kill about one hundred million sharks![8]

"Myth"

A myth is a report from the deeper layers of the world, which has been transmitted orally by Indigenous peoples over thousands of years. Creation myths do *not* relate to a distant past, but to the hidden dimension of causes behind the veils of the world of appearances. The myth of a people is a foundation of their relationship with the living Earth. Australian Aborigines wander once a year along the mythical songlines to sing the world into renewed existence, as did the Celtic bards in ancient Ireland. One aspect of true myths is how we can *serve* the world. Industrial society, on the other hand, uses the word as a synonym for superstition and deception, at best for "misunderstanding", at worst for "lies". This malicious distortion of the term is no accident: the myths of primitive peoples contain great strength to connect humans to the Earth, and therefore a great potential danger to the functioning of consumer society based on separation and alienation. Using a term that describes the sacred for Indigenous peoples as a synonym for lying is also an aggressive and deeply colonialist assault.

"Tipping point"

The latest case of word theft. The term tipping point stems from ecology and Earth system sciences and spread quickly through the 2019 climate debate in the wake of the IPCC report (released on October 8, 2018) and it's warning that humanity has only a decade left to tackle the global ecological crisis. Nobody knows (yet) when any of the planet's life support systems will tip over their tolerance limits. That would inevitably lead to global chain reaction beyond human control. In terms of climate breakdown, it could lead to "Hothouse Earth" and the annihilation of human civilization as we know it. In terms of biodiversity loss—for example, bees, whales, or plants—the outcome for us would be the same. *To consider, and fear, tipping points is the ultimate warning, and it should be treated so.* Instead, some politicians and mainstream media began in 2020 to diminish the term by using it out of context, like a mere substitute for "the last straw".

Footnotes

A. Hemerophiles (from Greek *hemeros* "cultivated" and *philos* "friend") are animals or plants that gain advantages from anthropogenic landscape changes, and therefore follow homo sapiens into the cultivated landscape (woodlands, fields, meadows, gardens, traffic routes, settlements, even dwellings). These include species such as field hare, field mouse, field hamster, partridge, quail, skylark, lapwing, carrion crow and cabbage butterfly. Closer to the buildings, house mouse, bats, sand martin, barn swallow, barn owl, and common redstart. Inside the buildings, housefly, spiders such as funnel weavers (*Tegenaria*) and great tremor or cellar spider (*Pholcus*), and silverfish. In urban areas city pigeon, swift, jackdaw, kestrel, and black redstart.

B. Here are more historical examples of ecocide: https://www.endecocide.org/en/examples-of-ecocide/

C. The first nations which have enacted ecocide in national law are Vietnam 1990, Belarus and Georgia 1999, Ukraine 2001, Republic of Moldova 2002, Armenia 2003, Ecuador 2008 (constitutional) and 2014 (criminal code), Uzbekistan 1994, Russian Federation 1996, Kazakhstan and Kyrgyzstan 1997, and Tajikistan 1998.[9]

Chapter 7

The Sixth Mass Extinction

The good news is, Earth has only three main problems: (plastic) pollution, mass extinction, and climate disruption. The bad news is, all three are enormous, and trying to tackle the causes humanity has gone round in circles for at least forty years.

The largest driver of biodiversity loss on land in recent decades has been "Land Use Change", primarily the conversion of pristine native habitats into agricultural systems to feed the world, while oceans are over-fished. This has been driven in large part by a doubling of the world's population, a fourfold increase in the global economy, and a tenfold increase in trade.[1]

Unlike climate disruption, mass extinction does not receive much media coverage and public discussion (apart from some articles about the disappearance of honeybees, native bird species, and such iconic animals as tigers and rhinos). But mass extinction is just as grave and just as menacing for planet Earth. One million species are threatened with extinction. And if they disappear the continued existence of the human species is also very much in question.[2]

"The air you breathe, the water you drink and the food you eat all rely on biodiversity," says Damian Carrington, environment editor of *The Guardian*, and continues with a quote from Prof. David Macdonald, Oxford University: "Without biodiversity, there is no future for humanity,"[3] because without it the ecosphere loses its basis for necessary adaptations to an ever-changing environment. The sixth mass extinction denotes nothing less than the worst loss of life on Earth since the demise of the dinosaurs about 66 million years ago. The full-scale eradication of vast populations of the planet's invertebrates, vertebrates, and plants. A "wave of *biological annihilation* that includes possible species extinctions on a mass scale, but also massive species die-offs and various kinds of massacres." (S. Banerjee)[4]

In each of Gaia's domains (sea, air, land) the symptoms of the gradual breakdown of nature's global life support systems are becoming overwhelmingly clear. Organisms suffer from (physical, chemical, nuclear, sonar, microwave) pollution, and with their habitats being destroyed (by expanding human settlement zones, industrial areas, resource extraction and transport routes) entire populations disappear at an alarming rate. Global heating[5] amplifies the speed of extinctions, and in turn the loss of biodiversity and the resulting weakening or even collapse of ecosystems accelerate climate breakdown.

The **loss of species diversity**, i.e. the number of species disappearing irrevocably, is alarming. And so is the speed with which it is happening: about 100 to 1,000 times faster than ever before in Earth's history.[A] "By contrast to this natural rate, thousands (if not tens of thousands) of species are vanishing yearly," says Prof. Eileen Crist.[6] One million species are threatened with extinction. "And the biosphere is not only hemorrhaging species, it is also losing its *abundance* of wilderness and wild creatures. The great masses of flocks, schools, and herds of animals are vanishing, and so are their migrations." *Habitat loss* is everywhere. Forests are disappearing, and half of the world's wetlands were lost in the twentieth century alone. Landscapes and seascapes are losing their ecological complexity, and the ensuing *simplification* of ecosystems is worsened "by globalization, that is swiftly homogenizing the biosphere." *Invasive species*, often migrating via human trade routes, threaten local ones, and in the ruptured landscapes "biodiversity—the unique loveliness of each place—first recedes, then vanishes."[7]

The naturalist Wade Davis reminds us of what abundance used to mean. About passenger pigeons: "In 1870, when their numbers were already greatly diminished, a single column one mile wide and 320 miles long, containing an estimated two billion birds, passed over Cincinnati on the Ohio River." Columns of pigeons blackened the sky, the noise of the birds taking flight was comparable "to that of a gale, the sound of their landing to thunder." And about fish: "Off the shores of Newfoundland, cod were so abundant that ships with wind in their sails made little progress, blocked by the density of fish in the water. Europe and much of the New World lived on the catch for 300 years. Then, in the years of my youth, factory ships industrialized the fishery and in a single generation reduced the species to a shadow in the sea."[8] It is becoming a world of shadows, the diminishment of life's richness is unraveling "species by species, population by population, habitat by habitat, and today (after the steady chiseling of centuries if not millennia) acre by acre." (Crist)[9]

The World Wildlife Fund's (WWF) Living Planet Report 2018[10] disclosed that a terrifying **60 percent** of the Earth's **mammals, birds, fish, and reptiles** have been lost since 1970.[B] The world's experts warn that the annihilation of wildlife is now an emergency that threatens civilization. According to a study published in the magazine *Nature,* marine **phytoplankton** in surface waters had declined by about **40 percent** from 1950 to 2010, possibly in response to ocean warming.[C,11] Populations of migratory river fish around the world have plunged by 76 percent since 1970, in Europe even 93 percent.[12] The Living Planet Report 2020[13] repeated the warning that global wildlife is in freefall. In only two years, global animal populations together had fallen from 40 percent to 32 percent (of their abundance in 1970).

On a healthy planet, **insects** abound on all continents. "At any one time, there are trillions of insects circulating in the air, known as 'bioflows', creating an intrinsically important dynamic ecological tapestry, where essential nutrients for ecosystems are continually circulated."[14] A long-term study in

Germany found that from 1990 to 2017, the abundance (biomass/ total weight) of flying insects decreased by **76 percent**. In other words, insect abundance has shrunk to less than a quarter.[D] 78 percent of wild plant species in the temperate zone and 94 percent in the tropics are pollinated by animals, not just by honeybees and wild bees, but also by flies, butterflies, moths, wasps, and beetles. *75 percent of the leading global food crops need pollination,* healthy populations of wild pollinators increase yields significantly.[15] (In 2019, pollinator loss is already putting up to $577bn (£440bn) of crop output at risk.)[16] "If we lose the insects then everything is going to collapse," says Prof. Dave Goulson of Sussex University, UK. "We are currently on course for ecological Armageddon."[17]

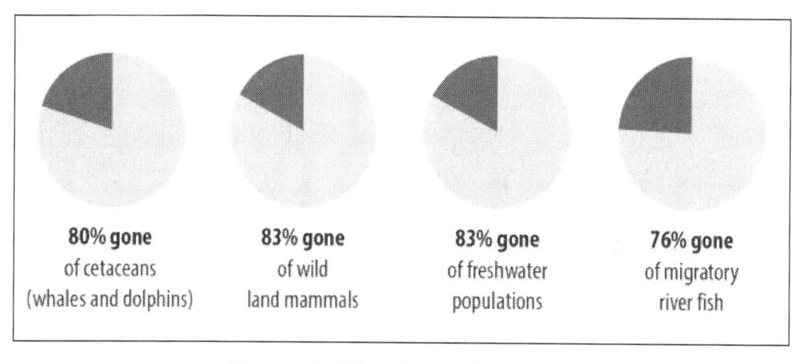

Figure 6. Wiped out since 1970

Trees are disappearing all over the world. Not just from vast-scale deforestation, but individually, tree by tree, species by species. In Africa, the ancient baobab trees, water-bearers for entire communities, are mysteriously dying.[18] In the Mediterranean, olive tree and date palm populations are seriously threatened by pandemics.[E] In northern Europe, larches, oaks, maples, chestnuts, and junipers are struggling with various pathogens.[F] Ash trees fall victim to ash dieback,[G] oaks to COD (Chronic oak dieback) and AOD (Acute oak decline).[19]

In the USA, brown needle blight has been found in 60 species of conifer so far.[20] Between 2001 and 2005, a tiny bark beetle (*Ips confusus*) killed more than 50 million piñon pine trees, about 90 percent of the mature ones in northern New Mexico. A number of other conifers are significantly declining across the USA,[H] but the biggest tree tragedy in modern America was the chestnut pandemic[I] which killed "an estimated three to four billion trees across more than 30 million acres."[21] According to the Forest Ecology and Management report 2010, many other tree species of the world are exiting their homelands, from Canada to Argentina, from Switzerland to Zimbabwe, from China to New Zealand.[22]

There have already been five global mass extinctions in the course of Earth's history (the last occurred 66 million years ago), and each time Gaia (over millions of years) produced a new and even greater biodiversity. But the notion that "Nature will bounce back alright" can not serve as justification for ecocide. Nature might not, this time. Gaia is strong, but there are warning signs: the fact that for the last 400,000 years the eccentricity of the Earth's orbit has regularly led to ice ages (compare box "Irregularities" at the end of Chapter 3, and Fig. 17) could well mean that Gaia's power of self-regulation is weakening. Can a living planet "age"?

What is being done?

The UN Convention on Biological Diversity has drafted a Paris-style UN agreement for the conservation and restoration of ecosystems and wildlife. To stop and reverse biodiversity decline, at least a third of the world's oceans and land need to be protected as soon as possible.[23] This needs all the support it can get. Unfortunately, UN resolutions are not binding, and we are witnessing year after year how the Paris climate agreement is far from making a real difference.

Let us look systematically at the causes of this global tragedy.

Harvard entomologist and biodiversity pioneer Edward O. Wilson[24] has summarized the multiple causes for biodiversity loss in the acronym HIPPO. This is not putting any blame on the good old hippopotamus. It's an abbreviation for **H**abitat destruction, **I**nvasive species, **P**ollution, **P**opulation, **O**verharvesting aka Overconsumption. Climate breakdown, while mainly being caused by the two Ps, is a huge driver for H.

Footnotes

A. Including the comet which 66 million years ago didn't kill the dinosaurs on first impact, but over the long period of climate changes that followed.

B. Population declines are particularly pronounced in the tropics: South and Central America, and the Caribbean, suffered the most dramatic decline with an 89 percent loss compared to 1970.[25]

C. An updated study revealed statistically significant chlorophyll declines (chlorophyll is a widely used proxy for phytoplankton biomass).[26]

D. There were even less insects, 18 percent, in the summer samples, when insect numbers should reach their peak. The fact that the samples were taken in nature reserves makes the findings even more worrying, because in the battlefields of agribusiness there is even less food and shelter for them.[27]

E. Olive trees die by the bacterium *Xyllela,* and date palms by the red palm weevil (Rhynchophorus ferrugineus).[28]

F. the fungus *Phytophtora*

G. by Chalara (*Hymenoscyphus* sp.)[29]

H. Other conifers declining are the black spruce, the white spruce, the ponderosa pine, the lodgepole pine, and the whitebark pine.

I. Britain in particular is also haunted by bleeding canker of horse chestnut (*Pseudomonas syringae pv. aesculi*).[30]

Chapter 8

Habitat Destruction

Human settlements have increasingly spread over the millennia, "widespread habitat destruction became a well-orchestrated global phenomenon only during the nineteenth century, with the onset of the Industrial Revolution." (Harding)[1] Before that, the Earth was still covered in a continuous mosaic of wild habitats that blended into each other, with only a minority of cultivated land.[A] Today, humanity claims almost every bit of land for food production, housing, industrial facilities, airports, harbors, traffic routes (roads, train tracks, waterways), power plants, and dams. For example, the number of dams has escalated in the past 50 years; worldwide there are now about 50,000 large dams (higher than 15 meters) and approximately 17 million reservoirs. They have flooded fertile river valley ecosystems of more than 154,000 square miles (400,000 square kilometers), destroyed habitats, and have displaced countless animals and 40-80 million humans, mostly poor farmers and Indigenous people. And as for transport routes, paved road lengths are projected to increase by 15.5 million miles (25 million kilometers) by 2050, ramping up the speed of human invasion and "resource" extraction.[2]

Agribusiness is the main driver for habitat loss and fragmentation (farmland currently makes up 45 percent of the EU's land area, and also in the USA). The trend to ever bigger monocultural fields crushes even the last biodiverse hedges and non-utilized strips of land. The plains of industrial farming with their pesticides are mass graveyards for insects and those who eat them (bats and birds accumulating toxins inside their bodies). In turn, the abundance of farmland birds in 28 European countries has fallen by 55 percent in the past three decades, according to the European Bird Census Council.[B] And bear in mind there was already a dramatic slump in bird populations in the 1960s when pesticides began their world-wide crusade.[3] And freshwater animal populations have collapsed by 81 percent since 1970,[4] following pollution, dams, and huge water extraction for agribusiness, industries and households.[C,5]

Trees in particular are affected by "climate-induced physiological stress" (storms and floods, erosion, severe drought, rapid warming). This is *global heating acting as a factor of habitat destruction*. The old habitats become too warm and dry, and individual trees, unlike fish, can not migrate fast enough to cooler regions, i.e. polewards. With the forests, all their biodiversity disappears too. However, the major factor for global loss of biodiversity and loss of biomass is not global heating yet, but deforestation, and it is steadily accelerating worldwide.

As the relentless destruction of the Amazon rainforest continues,[D] the world's largest forest, and the Earth's biggest terrestrial carbon sink, is now—significantly sooner than previously thought—getting close to a *tipping point* at which it could begin to tumble in a chain reaction and entirely disappear.[6] By 2020, as much as 40 percent of the existing Amazon rainforest received much less rain than they used to (due to global heating but also because years of large-scale deforestation changed the weather patterns of the Amazon basin). At the average rate of deforestation of the last few decades (let alone the accelerated destruction since 2018), the Amazon rainforest will have gone by about 2070, replaced by dry savannah. Only a thin belt of forest might remain in the Peruvian and Colombian Andes.[7]

In central India, the Hasdeo Arand with its 420,000 acres (170,000 hectares) is one of the largest continuous forests. It is rich in biodiversity and home to Indigenous peoples. Unfortunately it also sits on an estimated 5Gt of coal. Corona recovery India-style is to plan the opening of 40 new coalfields for commercial mining, regardless of the protest and objections from Indigenous villages and regional governments alike. India is already the world's second largest consumer of coal.[8]

We all know about the enormous destruction of forests for palm oil plantations in Indonesia where their surface had extended to more than 17 million acres (7 million hectares) by 2009.[9] And yet the ongoing palm-oil boom in industrialized countries continues to trigger more forest land conversions: West Kalimantan, Borneo, is in the process of assigning around 12 million acres (5 million hectares) of land—an area more than twice the size of Vermont—for oil palm plantations. The rest of the region is almost entirely reserved for other extractive industries, such as bauxite mining. Little land will be left for forests, or for local people to live and farm. These doomed regions in West Kalimantan consist of rainforests known for their orangutan populations and also of peatland which is especially rich in carbon. The destruction of Borneo's forests and peatlands contributes massively to GHG emissions worldwide, in some years making Indonesia the world's fourth-largest contributor to global heating.[10] And local and provincial Indonesian governments have plans to issue licenses for an additional 37 million acres (15 million hectares) of oil palm plantations.[11]

And, to name but two of many other examples, 1.2 million acres (half a million hectares) of Colombia's forest land, including tropical rainforests, have been cleared since 2006 for agricultural exploitation. The country doubled its number of palm oil plantations, and tripled exports to Europe.[12] The island of New Guinea, still home to large areas of intact tropical forest, is now under siege from the Trans-Papuan Highway[E] project, a 4,000km long "development corridor" for the extraction of "primary resources".

Think of all the other lives such as birds and animals, that found their homes and food around those trees. "What happened to them and how do we talk about that which we can't see and will never know?" (S. Banerjee)[13] Take the three to four billion (thousand million!) American chestnut trees whose abundance of flowers supported honeybees and other pollinators. Each autumn, as many as 6,000 sweet chestnuts per tree provided a wide spectrum of wildlife from turkeys to bears with vitamin C, protein, and carbohydrates.[14] A bird study on the Pajarito Plateau of New Mexico's Jemez Mountains, where some of the worst piñon pine die-offs occurred (between 2001 and 2005) shows that the diversity of birds declined by 45 percent, and bird abundance decreased by 73 percent.[15]

Another aspect of land grabbing and habitat destruction is the triggering and magnification of zoonotic diseases. Pathogens that have lived in animals in remote wilderness areas lose their hosts (because wildlife is being eradicated) and may jump onto domesticated animals, and eventually to humans. Pandemics such as SARS, MERS, West Nile, swine flu, and COVID-19 have their roots in habitat destruction (see "Pandemics", Chapter 9).

Fragmentation

Today, all habitats—forests, wetlands, tundra, mangrove swamps—have been seriously degraded and still face increasing threats. "Development" aka "resource" extraction such as deforestation and mining, road building, conversion of wild nature for agriculture, settlements, trade and industrial purposes continues unabashedly. Very few rivers, for example, still have flowing, healthy waters. Increasing numbers of dams, weirs, sluices and other barriers built in rivers are fragmenting the waterways, isolating habitats and weakening wildlife populations. In Europe alone, the number of water barriers has increased to 600,000, preventing, among many other things, the migration of salmon, sturgeon, and shad.[16] In Great Britain, 97 percent of the river network is fragmented; there is at least one artificial barrier every 1.5km (1640 yards) of stream.[17]

Fragmentation also happens in the sea. No less than two-thirds of the marine environment has been changed by fish farms, shipping routes, subsea mines and other projects.[18]

When humans attack and decimate the wilderness, they usually leave a few patches untouched. These become the last refuges for the remaining populations of the many wild creatures which once roamed so freely. But cornered in by inhospitable agricultural land, buildings and roads, these islands soon turn into concentration camps and hospices. Often, the fragments are too small to find suitable mates, enough food, or even a place to rest.

Example: In an intact Amazonian rainforest a wide variety of dung and carrion beetles of different sizes, shapes and specialized knowhow care for the swift recycling of all sorts of dung. As they feed precious nutrients to their offspring, they also eliminate parasites and bury seeds. But clearcut zones and forest fragments have fewer beetle species, sparser populations, and smaller beetles than comparable intact forest areas. The smaller the fragments the poorer the situation. *Edge effects* creep in: hot, dry winds from neighboring pastures or soy fields kill off dung beetle larvae; monkeys and birds that provided plenty of dung in the past have disappeared; and good mates for beetles are rare. The consequences of the dung beetle extinctions are more diseases among the few remaining birds and mammals, and soil nutrient loss due to erosion as heavy rains aren't caught anymore by plant roots because their seeds don't germinate anymore. The seemingly insignificant dung beetles turn out to be the *keystone species* of the forest.[19]

What is being done?

Creating nature reserves (as large as possible) and planning wildlife corridors between them is a major focus of the UN Convention on Biological Diversity (CBD), many governmental agencies, and NGOs. The 2018 UN Biodiversity Conference of the Parties (COP14) closed with broad international agreement on reversing the global destruction of nature and biodiversity loss that is threatening all forms of life on Earth, and with high hopes for an ambitious new global agreement at the follow-up conference in Beijing in 2021. So much for the dreaming.

In reality, it is different. The world failed to meet a single one of the twenty UN biodiversity targets agreed in Japan 2010 to be achieved by 2020. The UN Decade on Biodiversity was a lost decade for biodiversity. With the disappointment of 2020, the CBD began to prepare targets for 2050.[20] In general, the biodiversity conferences have been receiving miserably little attention, "the spirit of international collaboration appears to be as much at risk of extinction as the world's endangered wildlife." (*The Guardian*)[21] Heads of state don't show, and even environmental ministers aren't a given; Germany, for example, sent only an undersecretary of state in 2018. As the target to expand nature reserves from 10 percent (in 2010) to 17 percent (by 2020) of the world's land failed, Pooven Moodley director of the NGO Natural Justice said, "It does feel like rearranging the deckchairs as the Titanic sinks."[22]

At this rate humanity could be the first species to document its own extinction, warns CBD executive secretary Cristiana Pașca Palmer. "The loss of biodiversity is a silent killer," she says. "It's different from climate change, where people feel the impact in everyday life. With biodiversity, it is not so clear but by the time you feel what is happening, it may be too late."[23]

There are some glimmers of hope, of course. In Canada, where First Nations are constitutionally guaranteed the right to harvest salmon from the rivers, numerous Indigenous communities along the massive Fraser and Skeena Rivers have volunteered to dramatically scale back, if not halt, their annual salmon harvest.[24] And in June 2018, the Spanish government created a Mediterranean Sea reserve for whale migration.

Many conservationists argue that protecting 17 percent of the world's land will still not halt extinctions. Conservation targets should not be determined by the little that humans *are willing* to spare, but by what *is necessary* to protect nature. The organization Nature Needs Half (natureneedshalf.org) takes a far bolder approach and campaigns for the preservation of a full 50 percent of our planet for nature by 2050. The idea is supported by the distinguished Harvard biologist EO Wilson, who even named his most recent book, *Half-Earth*,[25] accordingly. "We thrash about, appallingly led, with no particular goal other than economic growth and unfettered consumption," he writes.[26]

The EO Wilson Biodiversity Foundation started The Half-Earth Project as a "call to protect half the land and sea in order to manage sufficient habitat to reverse the species extinction crisis and ensure the long-term health of our planet."[27]

A meaningful map identifying Earth's regions to save biodiversity, the last intact ecosystems on the planet—and the climate—has been constructed by a team of scientists and published in *Science Advances* in September 2020. This "Global Safety Net" would increase the protected land area from currently 15 percent to 50.4 percent of Earth's land mass.[28] Half Earth.

On similar grounds, an emerging global **rewilding** movement has spawned the Global Charter for Rewilding the Earth. Working together with "the broadest spectrum of constituencies and encouraged by governmental policy [the vision is to] ultimately weave wondrous blue and green ribbons of wildness that wrap the Earth in beauty, offering the promise of a better future, with freedom and habitat for all."[29]

The principles for rewilding:

- The ecosphere is based on relationships

- Making hopeful stories come to life
- Embracing natural solutions and thinking creatively
- Protecting the best, rewilding the rest
- Letting nature lead
- Working at nature's scale
- Taking the long view
- Building local economies
- Recalling ecological history and acting in context
- Evidence-based adaptive management
- Public/private collaboration
- Working together for the good of ourselves and nature.

What can I do?

- Spread the word: Talk to friends and colleagues about the importance of biodiversity. Sign petitions and/or write your own messages to (local, regional, national) politicians.
- Support EO Wilson's Half-Earth Project: half-earthproject.org and/or Nature Needs Half: natureneedshalf.org and
- If you think 50 percent is too ambitious at this stage of human (under)development, go for the 30 percent campaign: campaignfornature.org/petition
- Help rewilding in your area.
- Campaign against pesticides, neonicotinoids, etc.
- Support the Greenpeace Global Oceans Campaign,[30] and others.
- With or without children, watch the short Greenpeace video "Rang-Tan in my bedroom".[31]
- If you have a garden or balcony, feed the birds in winter and spring (unless you have a cat). But instead of dirty birdhouses use more hygienic bird feeding silos.

Footnotes

A. After the last Ice Age (some 10,000 years ago), we could have walked from Western Europe to the great rainforests of Thailand and Vietnam without ever leaving nature's wild habitats. "The abundance of flying, leaping, and swimming creatures in this pristine state astonished the first European settlers all over the world, who quickly set about logging, hunting, fishing, and clearing for agriculture with a demonic destructiveness that beggars the imagination." (Harding)[32]

B. A disturbing decline in French farmland birds was revealed by two studies in March 2018. Bird populations in France have fallen by an average of a third over the past 15 years. "We are turning our farmland into a desert. We are losing everything and we need that nature, that biodiversity—agriculture needs pollinators and the soil fauna. Without that, ultimately, we will die," says Dr. Benoit Fontaine of France's National Museum of Natural History.

C. Wetlands are the habitats most impacted by human colonization, having lost 87 percent of their extent in the modern era.[33]

Canadian cod fisheries, like many other fisheries around the world, have been severely depleted by decades of *overfishing* and *short-sighted mismanagement*, but also poor *logging practices* and *dam construction*. "In some rivers, we've seen the numbers drop from the millions to the thousands," says Gerald Michel of the Xwisten Nation in central British Columbia.[34]

D. Between August 2017 and July 2018, 3,000 square miles (7,900 square kilometers) were deforested in the Brazilian Amazon rainforest alone: "a 13.7 percent rise on the previous year and the biggest area of forest cleared since 2008. The area is equivalent to 987,000 football pitches."[35] Twelve months later, the area of newly destroyed forest was almost 3770 square miles (10,000 square kilometers).[36]

E. The "Trans-Papuan Highway" is a project by the Indonesian federal government. In the Papua and West Papua provinces, the new roads will cut through the traditional lands of many different Indigenous groups, and also through the Lorentz World Heritage Site.[37]

Chapter 9

Invasive Species

Invasive species—mostly spreading in the wake of human activities—are one of the five top-drivers for biodiversity loss. Even in comparatively healthy ecosystems with little habitat fragmentation, they can cause considerable damage. And the problem is increasing due to tourism and globalization.[A]

Most introduced species don't do harm, they blend in or fade out again. But a minority take hold and cause massive disturbance. The absence of their own natural predators and diseases gives them an advantage over the native species. Pests and parasites migrate too. Mosquitos venture more north each year; so do ticks, the second-most-dangerous carriers of infectious diseases. In Central Europe, new and dangerous ticks from Africa have been arriving since 2015, and spread further north year by year.[1]

The ethical question about whether or not to try and control (cull) invasive species is to do with *animal ethics* but also with *environmental ethics*. It's not about pitching the native against the exotic species, but about the whole ecosystem of which they are (or not) a part.

Example: In Europe, the red squirrel (*Sciurus vulgaris*) has been native since time immemorial. It even has a place in Nordic mythology where it runs up and down the World Tree as a messenger between the divine forces above and the divine forces below (underworld). But since the American grey squirrel (*Sciurus carolinensis*) was introduced to the British Isles in 1889, it has pushed out the red squirrel in almost all areas. The main reason for this is the squirrelpox virus to which the grey is immune but the red is not. However, the "replacement" of the red with the grey squirrel affects the entire ecosystems.

Song bird numbers are dropping in "grey" areas, due to food competition and because the greys occasionally rob bird's nests (they also eat insects, and frogs sometimes). And grey squirrels eat tree bark. Europe has many trees which are particularly susceptible to bark attacks, like beech, birch, yew, maple, and fruit trees such as cherry. Hence, not only human beings have a say in the matter of whether the grey squirrel should be left on its course ("They are 'nature' too, aren't they?"); the birds, the frogs and the trees vote for the return of the red.

What is being done?

There are programs to monitor and control invasive species.[B] Killing animals is not the only way; non-lethal control methods include reproduction limitation, and the impact of natural predators. For example, studies from Ireland show that the red squirrel has an old ally: the native pine marten. The grey squirrels are bigger than the reds and not as swift; to them the marten is a serious threat.[2]

For people who work in conservation, there are seven principles to assist the ethical evaluation of any proposal for invasive animal eradication:

Seven principles for guiding management of human-wildlife conflicts, ***to be followed in sequence*** (Dubois et al., 2017)[3]

1. When possible, modify human practices first.
2. Justify the need for control.
3. Have clear and achievable outcome-based objectives.
4. Cause the least harm to the fewest number of animals.
5. Consider community values as well as scientific and technical information.
6. Include long-term systematic management plans.
7. Base control on specifics of the situation rather than negative labels applied to the target species in question.

What can I do?

- As a foreign traveler, don't bring animals, plants or seeds back home. Ever.
- If you have animals in your care, follow the **Four guiding principles of animal ethics** (Fraser 2012)[4]
  1. Provide good lives for the animals in your care.
  2. Treat suffering with compassion.
  3. Be mindful of unseen harm.
  4. Protect the life-sustaining processes and balances of nature.

Mass Mortality Events (MMEs)

An entirely different type of "invasion" are the mass annihilations that seem to occur out of nowhere, and have been dubbed Mass Mortality Events (MMEs).

In spring 2015 in Kazakhstan (Asia), an estimated 200,000 saiga antelopes (*Saiga tatarica*) were grazing, loosely scattered over an area of 7.7 square miles (20 square kilometers). But within two to three days they all fell ill, and by the end of that week, every single one was dead. The puzzling thing was that this was not even a pandemic: there was no time for transmission from animal to animal. After 32 postmortems, the mysterious reason was found. Indeed there was no invading pathogen from outside, but a bacterium (*Pasteurella multocida*) which normally lives harmlessly in the tonsils of the antelopes. But a spring heatwave with 98.6 Fahrenheit (37°C) and an increase in humidity above 80 percent had stimulated the bacteria to overpopulate and pass into the bloodstream of its host animals where it caused blood poisoning (hemorrhagic septicemia).[5]

A mass mortality event (MME) is a single, catastrophic incident that wipes out vast numbers of a species in very short time. MMEs are among the most extreme phenomena of nature, and can push species to the brink of extinction. They are on the rise and likely to become more common because of climate change. Another huge MME triggered by temperature occurred in 2013, when the American west coast from Mexico to Alaska saw the biggest die-offs ever observed in the natural world. Hundreds of millions of starfish—more than 20 species—began to "melt" into white gloop. Again, similar to the MME in saiga antelopes, the starfish virus had been present in starfish for decades before the disaster. But the warming of the Pacific Ocean, induced by anthropogenic climate breakdown, had put stress on the animals while it made the virus more virulent.[6]

Other MMEs include:

- In 2013, another mass die-off happened at the edge of the same waters, along the Pacific coast of the US and Canada. It seemed to be "one of the largest mass die-offs of seabirds ever recorded" (wrote Craig Welch in *National Geographic* in 2015). And many more seabirds have been dying ever since (including Cassin's auklets, thick-billed murres, common murres, fork-tailed petrels, short-tailed shearwaters, black-legged kittiwakes, and northern fulmars).[7]
- In January 2018, temperatures in Sydney topped 116 Fahrenheit (47°C), and corpses of critically endangered flying foxes began to pile up under the trees in New South Wales. In Campbelltown, south of Sydney, 400 bats died in one spot.[8]
- In autumn 2018, scientists could finally identify the pathogen for the horrible "proliferate darkening syndrome" in river trout (*Salmo trutta fario*) in the Alps. The PRV virus can also be found in trout in Japan, USA, Canada, and Norway, and in salmon in Chile. Perhaps these MMEs can be prevented.[9]
- In autumn 2020, hundreds of thousands of migratory birds—flycatchers, swallows, and warblers—literally fell

> out of the sky dead. The had left their northern domains too early because of disrupted weather patterns, and hence didn't have enough fat reserves for their long journey to Central America. Then they met the historic wildfires across California, Oregon and Washington. No insects to feed on, and instead toxic smoke for their little lungs.[10]

Humans are beginning to feel the heat too. MMEs also raise the specter of climate change in terms of melting permafrost soils in the Arctic regions. In August 2016, a boy died of anthrax in the remote Yamal Peninsula, and 20 other people got infected but were treated and survived. Anthrax hadn't been seen in the region for 75 years. The outbreak followed an intense heatwave in Siberia, with temperatures over 30°C melting the frozen permafrost. "Long dormant spores of the highly infectious anthrax bacteria frozen in the carcass of an infected reindeer rejuvenated themselves and infected herds of reindeer and eventually local people," explains Jeremy Plester in his *Guardian* article "All hell breaks loose as the tundra thaws."[11]

Pandemics

Among the human species, MMEs have been known throughout history, they are usually called epidemics (at the level of a region or community) or pandemics (at the level of an entire country, continent, or the whole world). Ever since the Spanish flu, also known as the 1918 flu pandemic, killed an estimated 50 million to 100 million people worldwide, fears of another major pandemic have been present. Increasingly so since the turn of the millennium because human invasion of even the last refuges of wild nature releases ever-new pathogens and zoonotic diseases.

It is this ecocide that constantly releases new viruses and pathogens—such as SARS-CoV-2, the pathogen that causes

COVID-19—and it is the globalization of travel and trade that supports the ultra-fast spread of new diseases. "AIDS, Ebola, West Nile, SARS, Lyme disease and hundreds more that have occurred over the last several decades—don't just happen. They are a result of things people do to nature," said Jim Robbins in *The New York Times*, already in 2012.[12]

A few examples:

- AIDS crossed into humans from chimpanzees in the 1920s when bush-meat hunters in Africa killed and butchered them.[13]
- In 1999 the nipah virus raged in South Asia, after an infected flying fox (*Pteropus vampyrus*) dropped a piece of chewed fruit into a piggery in a forest in rural Malaysia. The pigs became infected with the virus and amplified it. It jumped to humans where it revealed a horrifying mortality rate of 38 percent.[14]
- In 2002/2003, the outbreak of SARS (Severe acute respiratory syndrome) began with cave-dwelling horseshoe bats and a farmer in China's Guangdong Province.
- In the Amazon, the human invasion of intact tropical landscapes shows time and again that disease follows deforestation. It has been shown that an increase in deforestation by just over 4 percent increases the spread of malaria by nearly 50 percent. This is because mosquitoes, which transmit the disease, thrive in the right mix of sunlight and water in recently deforested areas (rainwater is not taken up by tree roots anymore, and the tire marks of the heavy machinery leave compressed soil with myriads of puddles in which mosquitoes thrive).[15]
- Lyme disease is the result of the reduction and fragmentation of large continuous forests. Human development chased off the predators—wolves, foxes,

owls and hawks—leaving the species which are the greatest "reservoirs" for pathogens. On the US East Coast it's white-footed mice, across Europe it's deer. A single deer can carry thousands of tick eggs. And ticks bring not only Lyme disease but others too (e.g. babesiosis and anaplasmosis).[16] And as a result of climate disruption, other tick species have been coming to Europe from Africa since 2015.[17]

- SARS-CoV-2 is believed to have begun its spread around the globe in a Chinese "wet market" for wildlife meat in late 2019. Again, the virus originated from bats which should have been left alone in their remote cave dwellings.

What all these case histories (and many others, such as swine flu and MERS) have in common is that they began at the frontline, either where humans brutally intervene in remaining ecosystems (poaching, deforestation, road construction, the development of mines, dams, etc.), or where, due to overpopulation and urbanization, land grabbing pushes into the last refuges of wild nature.

"Any emerging disease in the last 30 or 40 years has come about as a result of encroachment into wild lands and changes in demography," says Peter Daszak, a disease ecologist and the president of EcoHealth.[18] This does not exclude the threat of animal husbandry, on the contrary. Domesticated animal species have been carriers of pathogens since the beginning of the history of civilization, some of which can also be transmitted to humans.[C] As a result of increasing globalization—increased international transport of meat and live animals—germs of the most diverse origins meet in countless locations and can form new crosses and mutations.

Today the threat of pandemics is greater than ever before because a) *global habitat destruction* is fiercely accelerating, b) *population density* of domestic animals as well as humans means that outbreaks which would have affected only small local groups in the past are now amplified and spread fast in overcrowded livestock confines and human settlements, and c) *globalization* with its high-frequency network of trade routes and cheap travel distributes germs faster than emergency shutdowns can come into force.

The Western world at large was complacently content to blame the Asian wet markets for COVID-19, but the livestock trade and industrialized abattoirs are just as unsafe. In 2020, the COVID-19 crisis exposed not only the unspeakable levels of cruelty towards animals in industrial meat production but also the appalling and sickening work conditions for the workers who execute the daily bloodbaths.

The "free flow of goods and people" so hailed by neoliberal globalists inevitably is also a free flow of pathogens. Almost 30 percent more pigs, goats, cows and sheep were shipped, flown and driven across the world in 2017 than a decade earlier, according to FAO (the UN's Food and Agriculture Organization). As in the Malaysian outbreak of nipah mentioned above, the progression of the African swine fever virus (ASF) into Europe, and the spreads of avian influenza virus and mad cow disease were assisted by the livestock trade.[19]

The globalized livestock and meat trade is irresponsible and unsustainable. The danger of pandemics is constantly growing as we drive the planet's life support systems closer to the abyss. Profit-driven meat production "offers the exact means by which pathogens can evolve the most virulent and infectious phenotypes," says biologist Rob Wallace, author of *Big Farms Make Big Flu*. "You couldn't design a better system to breed deadly diseases."[20]

What is being done?

Next to nothing. The broad-scale land grabbing and unabashed "resource" extraction continue. Humanity is the most invasive species of all. Rob Wallace emphasizes that "anyone who aims to understand why viruses are becoming more dangerous must investigate the industrial model of agriculture and, more specifically, livestock production. At present, few governments, and few scientists, are prepared to do so."[21]

Global nonprofit environmental health organizations like the One Health Initiative[D] and the EcoHealth Alliance[E] and governmental programs like Predict[F] merely deal with the pathogenic symptoms of planetary sickness, not the root causes.

Footnotes

A. Introduced species include domestic animals (cattle, pigs, goats, cats, rabbits, etc.) that were brought along on purpose, and accidental travelers like the bug that caused the Dutch elm disease which spread all around the world via infected timber imports and destroyed millions of elm trees in two pandemic waves in the twentieth century. In the USA, 4,000 exotic plant species and 2,300 exotic animal species have been introduced so far, and they are threatening 42 percent of the native species on the endangered species list.

B. In early 2018, The Global Registry of Introduced and Invasive Species was launched, to catalogue these and stand alongside the IUCN Red List of Threatened Species as an international means to fight extinction, by helping to stop biological invasions.[22]

C. Humanity acquired almost all of its infectious diseases from animals, already in ancient times: e.g. whooping cough, meningitis, diphtheria, polio, smallpox, measles, mumps, infectious hepatitis.[23]

D. The One Health Initiative "seeks to promote, improve, and defend the health and well-being of all species by enhancing cooperation and collaboration between physicians, veterinarians, other scientific health and environmental professionals." It is supported by dozens of organizations, and during the first COVID-19 lockdown the number of participating scientists in these fields rose from 600 to over 950. (onehealthinitiative.com)

E. The EcoHealth Alliance is a global nonprofit environmental health organization that partners with more than 30 countries worldwide. "The urgent concern for wildlife conservation and the overall health of our planet has led EcoHealth Alliance to become an environmental science and public health leader working to prevent pandemics in global hotspot regions across the globe and to promote conservation" (ecohealthalliance.org).

F. PREDICT was a prevention program initiated in 2009 by the US Agency for International Development to strengthen global capacity for detection and discovery of zoonotic pathogens with pandemic potential. It was discontinued by the US government in October 2019, ironically two months before the outbreak of COVID-19. During its two decades, it identified at least 931 novel virus species (!) from samples of wildlife, livestock, and humans. (Among them, hundreds of different corona species.)[24]

Chapter 10

Pollution

Plastics

Contamination with plastics is by now spanning the entire globe. Microplastics are found everywhere from high mountains to deep oceans. They travel extensively by wind and via waterways and reach each living organism on Earth. In 2019, meltwater in the Arctic contained an average of 1,760 microplastic particles per liter, meltwater in the Alps even 24,600 particles per liter.[1]

We breathe in synthetic fibers (already in 1998 they have been found in 87 percent of human lung tissue samples, and were present in 97 percent of malignant lung tumors[2]), we drink them (about 4,000 microplastics particles per year, and an additional 5,000 particles if the water comes from plastic bottles[3]), and we eat them, from 39,000 to 52,000 particles per year, depending on age and sex.[4] This is a lot, but nothing in comparison to what bottle-fed babies have to put up with. Because the recommended heating process for sterilizing polypropylene bottles and preparing formula milk causes the bottles to shed millions of microplastics and trillions of even smaller nanoplastics, babies in the USA swallow 2.5 million microplastic particles a day, babies in the UK 3.1 million.[5]

Worldwide, about 8,300 million metric tons (Mt) of plastics were produced until 2015. A part is still in use, but the bulk is our heritage so far: approximately 6,300Mt of plastic waste (around 9 percent of which had been recycled, 12 percent was incinerated, and 79 percent ended in landfills or elsewhere in the natural environment). If business-as-usual should continue, this number will almost double by 2050.[6] Current annual production of about 300Mt new plastics is expected to steeply rise soon (see below), and to reach nearly 2,000Mt annually by 2050.[7] And the plastic waste reaching the oceans is even expected to nearly triple until 2040, by then adding 29Mt a year, the equivalent of 50kg for every meter of coastline in the world.[8]

Greater use of plastics also results in more extensive consumption of fossil fuels and, in turn, in larger emissions of greenhouse gases, including methane.

A breakdown of the biggest sources of plastic pollution (other than industrial fishing gear)

The UK as a typical industrial nation (in 2017):[9]

- Vehicle tires: 68,000 metric tons of microplastics from tire tread abrasion are generated every year, 7,000-19,000 metric tons of which enter surface waters;
- Clothing: washing synthetic clothing may release 2,300-5,900 metric tons of microfibers every year, up to 2,900 metric tons of which cannot be filtered by wastewater treatment facilities and reach rivers and estuaries;
- Plastic pellets used to manufacture plastic items: up to 5,900 metric tons are lost to surface waters every year;
- Paints on buildings and road markings: weathering and flake-off result in 1,400-3,700 metric tons ending up in surface water every year.

Tires seem to be the biggest contributor. They wear down through friction and tiny particles become airborne. In Europe, this produces 500,000 metric tons of plastic particles annually. Global emissions of tire wear particles (TWPs) amount to 6.1m metric tons, making them *the second-largest microplastic pollutant in the oceans after single-use plastic*. In road traffic, emissions of particles from brake linings, a complicated mixture of metal and plastic, add another 0.5 million metric tons worldwide. A recent study by the Norwegian Institute for Air Research[10] revealed that more than 200,000 tons of tiny plastic particles are blown from roads into the oceans every year. This could get worse with the spread of electric vehicles because due to their batteries they tend to be heavier than comparable gasoline or diesel cars, resulting in more wear on tires and brakes. Furthermore, since these particles are colored, they absorb light and, together with the black carbon in soot and particulates from burning fossil fuels, decrease the surface albedo of the planet, increasing global heating.

About a fifth of the plastic waste found in the oceans comes via rivers. Over 90 percent of the plastic debris that reaches the open sea comes from only ten rivers, eight in Asia and two in Africa.[11] However, don't blame these continents: they import vast amounts of waste from Europe and North America.[A]

In August 2018, around 300 endangered olive ridley turtles (*Lepidochelys olivacea*) were found dead off the southern coast of Mexico, trapped in an abandoned fishing net. The deaths of the sea turtles tragically points out the dangers posed by lost or discarded fishing equipment, so-called ghost gear. According to the UN Environment Programme (UNEP), "each year more than 100,000 whales, dolphins, seals and turtles get caught in abandoned or lost fishing nets, long lines, fish traps and lobster pots. Some of the abandoned nets can be as big as football pitches, and this plastic-based ghost gear can take up to 600 years to break down, shedding microplastics as it degrades."[12]

Microplastic particles (measuring less than 0.5cm = 0.2in) are the result of all sorts of plastic items which weather on land or disintegrate into ever smaller pieces aided by the mechanical forces of ocean waves. But they too are not the end of the chain: disintegration continues to the nano size of single molecules which contaminate microorganisms, accumulate upwards through the food chain, and poison all biota along the way. Also, due to their chemical and physical characteristics, micro- and **nanoplastics** attract other chemical toxins which bind to them. This increases their threat to organisms.

A special form of microplastics are **microfibers** which stem from synthetic clothing and carpeting. Each cycle of a washing machine can release more than 700,000 microscopic plastic fibers into the environment when you wash synthetic clothes. Acrylic is the worst, releasing nearly 730,000 particles per wash, nearly 1.5 times as many as polyester, and five times more than polyester-cotton blend fabric.[13] It's not just all fishing nets and plastic bottles. Microfibers shed by the everyday wear and tear of synthetic clothes and carpets contaminate the air in our buildings and eventually the atmosphere. Tumble dryers usually vent to the open air. These cumulative atmospheric inputs add to the atmospheric fallout which eventually contaminates the soil and the water.[14] Microfibers poison the different food chains, build up in animals' digestive tracts, and reduce their ability to absorb energy from foods in the normal way. Crabs that ingested food containing polypropylene rope microfibers (up to 1 percent plastic by weight) showed reduced food consumption and a significant reduction in energy available for growth.[15] Chinook salmon on the west coast of Canada often weigh only half as much as they used to.[16]

A third group are **microbeads** in cosmetics. In the UK alone, about 86 metric tons of microbeads were released into nature's cycles every year just from facial exfoliants.[17] Microbeads too get eaten by marine life, and have been shown to stunt growth, alter behavior, and even kill fish before they reach reproductive age.[18]

Microfibers and microbeads have been tiny to begin with, while microplastics in the strict sense started out bigger but have been pulverized by the ocean waves. But they usually all come under the umbrella of "microplastics". Microplastics have been accumulating in nature since the 1960s. *Microplastics are now the most abundant form of solid-waste pollution on Earth.*

While plastic garbage on land simply "disappears" from our view by becoming part of the layers of soil, it also sinks to the bottom of the oceans, but a big part gathers in surface waters. Since about 2010, people finally began to notice, hearing about the accumulated plastic trash in the Pacific known as the Great Pacific Garbage Patch. The "patch" covers an area of over 617,000 square miles (1.6 million square kilometers)—four times the size of California. It contains at least 79,000 metric tons of plastic. Microplastics make up the majority of the estimated 1.8 trillion pieces floating in the garbage patch, while nearly half of the weight of rubbish is composed of discarded fishing nets. Other items include buoys and ropes, bottles, plates, and the whole enchilada of household "goods".[19] And there are other such "islands" in the other oceans.

What eventually sinks can go deep—right down into the ocean trenches which are the deepest places on the Earth's surface. Crustaceans found in the deep trenches of the Pacific Ocean—these locations range from 4 to more than 6 miles (7-10 kilometers) deep—are also contaminated.[20] The microplastic concentration in those deep waters increased with depth, from 13 pieces per liter in water to a maximum of 2,200 pieces per liter in the sediments on the sea floor.[21] Humanity has truly hit rock bottom.

It's not just the high sea. First evidence of plastic contamination in freshwater fish surfaced in late 2018 in Brazil. Plastic particles were found in 80 percent of the species examined.[B] Microplastics have also been found in insects: in half of the mayfly and caddisfly larvae in rivers in Wales, and in mosquitos around the world which absorbed them as larvae in polluted waters, now contaminating new environments and threatening birds and other animals that feed on the insects.[22]

These new analyses made clear the extent of microplastic contamination in all of Gaia's domains. Already in 2016, Frank Kelly, professor of environmental health at King's College London, told a UK parliamentary inquiry about microfibers: "If we breathe them in they could potentially deliver chemicals to the lower parts of our lungs and maybe even across into our circulation." Dr. Anne Marie Mahon at the Galway-Mayo Institute of Technology, expands on the effects of microplastic contamination on living organisms: "If the fibers are there, it is possible that the nanoparticles are there too that we can't measure. Once they are in the nanometer range they can really penetrate a cell and that means they can penetrate organs, and that would be worrying."[23]

A completely unexpected discovery was published in July 2020: small crustaceans can break down microplastics into nanoplastic pieces, within just 96 hours. The study from University College Cork (UCC) in Ireland showed that microbeads of polyethylene were broken down by the crustaceans, the 2cm-long amphipod *Gammarus duebeni*, into nanoplastics that measured less than one micron, or one thousandth of a millimeter. If this biological fragmentation should also be found in other crustacean species, and in large populations, it would need to be considered a global issue.[24]

There is no end in sight yet. In May 2017, Exxon Mobil signed a $10bn agreement with Saudi Arabia to build the world's largest plastics facility on the Texas coast. And it is just one of 11 chemical, refining, lubricant and gas projects Exxon is building in the region. Over the next decade, this boom will fuel an anticipated 40 percent rise in global plastic production. With "efforts" like this, the amount of plastic in the oceans could even treble within the next decade, according to a UK government report.[25]

For years we have seen heart-breaking images of turtles suffocating from six-pack rings from drink cans stuck around their necks, lizards caught in plastic nets, and toothbrushes and golf balls from the stomachs of seabirds—by now, 90 percent of seabird species have plastic in their stomachs.[26] But since the human beast is capable of denial and looking the other way, these stark images of animal torture have made no considerable difference to the course of human affairs. Until September 2017, that is, when the plastic wave hit home true and hard. Tap water samples from more than a dozen nations were analyzed by scientists who found that 83 percent of the samples were contaminated with plastic fibers. The highest contamination rate was found in the USA, with 94 percent of the samples being contaminated. Water in Lebanon and India is not far behind, the best results were found in European nations (including the UK, Germany and France), still at 72 percent. Subsequent studies found plastic particles in sea salt, beer, honey and sugar.

No surprise then, microplastics are found in human excretions too: a small pilot study by the Medical University of Vienna analyzed the stool samples of eight probands (from Finland, the Netherlands, the UK, Italy, Poland, Russia, Japan, and Austria). All of them tested positive, with a total of nine different types of microplastic (Umweltbundesamt 2018). Similar finds in human kidneys and urine samples led medical scientists to doubt if they can ever run large studies about the health impact of microplastics. The reason: With all humans being contaminated already, one can't recruit a plastic-free comparison group anywhere.[27]

However, what *leaves* the body is not the biggest health problem, although it takes precious energy and immune system resources to dispel it. (More on health in the next section.) Microplastics have also been found in seabirds, fish and whales, because the animals swallow but cannot digest them, which leads to a buildup in their digestive tracts.[28] While toothed whales starve because too many big plastic pieces occupy their stomachs, filter-feeders such as baleen whales and basking sharks, which feed through filtering seawater for plankton, suffer from accumulating microplastics in their guts.[29]

Another dire example are PCBs (polychlorinated biphenyls). Having accumulated in the oceans and in the marine food chains they now threaten half the world's orca whale (*Orcinus orca*) populations. PCBs damage their immune system and impair their reproduction. It is particularly tragic that the fat-rich milk of the mother whales passes on very high amounts of PCBs to their newborn calves. The future of over half of the surviving orca populations worldwide is threatened by this.[30] And orcas[C] don't have an easy time anyway, since the salmon they feed on have largely diminished in size and numbers.[31] Meanwhile, female seals in the Baltic Sea were found with a constriction or closure of the womb, due to high PCB contamination. The otter decline across Europe has also been attributed to PCB. The stable otter populations in Britain survive with below 50mg/kg PCB in their liver fat.[32] All this is all the more shocking since PCBs have been banned since 2004 because of their toxicity.

What is being done?

Awareness of plastic pollution has increased in diplomatic circles. Almost 200 countries signed a UN resolution in 2017 that aims to stem the flood of plastic into the oceans. But as so often (sigh!), the agreement has no timetable and is not legally binding.[33]

Plastic bags seem to be at the forefront of political focus, and as of 1 January 2020, bans against them have been introduced in 74 countries, with varying degrees of enforcement, and 37 further countries have imposed a charge per bag.[34] The UK is in the latter group and also banned single-use plastic straws and cutlery, stirrers and cotton buds in 2020. The EU followed in 2021.[35] In the USA, California and Hawaii were the first states to ban straws, etc., already in 2018. But the nation is hesitating, and Florida even ponders over a bill to ban possible plastic straw bans until 2024.[36]

And what about plastic bottles? Synthetic clothing and synthetic carpets still going to landfill... or into the sea? With 4.7 billion plastic straws and 316 million plastic stirrers in the UK alone,[37] banning these single-use items is admittedly a great step forward, but still only a drop in the ocean. Considering the unimaginable gravity that by now every single ecosystem and every single organism on Earth is deeply contaminated with micro- and nanoplastics, political measures so far are borderline caricature.

Especially so when, after two years of mild public discussions about reducing plastic consumption, the earliest COVID-19 measures comprised of the idea to put synthetic masks and plastic gloves on everyone. France alone ordered 2 billion masks in April 2020, to begin with.[38] Already one month later, used disposable masks and latex gloves began to show up on shorelines around the world.[39]

A global deluge of single-use items each of which has a lifespan of 450 years... in the soil or the sea. Yet another timebomb: in June 2020, the World Health Organization (WHO) advised that everybody should wear a three-layer mask, with the two outer layers being made of polypropylene.[40] In theory, this plastic type is recyclable—but which countries are actually considering to set up adequate disposal channels?

One ray of hope: in the autumn of 2020 a new British standard for biodegradable plastic was introduced. Developed by a British company in cooperation with Imperial College London, and agreed by the Department for Environment, Food and Rural Affairs (DEFRA), it is set to ensure that biodegradable plastics break down within two years to nothing but carbon dioxide, water and sludge, without shedding microplastics or nanoplastics along the way. To meet the standard, plastics will have to pass tests which show they will biodegrade to a harmless state in real-world situations.[41] Such plastics could be used for the most common litter items such as food cartons, cling films, and bottles. However, CO_2 from plastic, just like CO_2 from the mineral oil it was made from, still adds to the carbon footprint.

Apart from promises, nothing is changing. Production levels are set to rise, feeding profits and global GDP while poisoning the planet. All measures listed below as positive achievements try to ease the symptoms but nothing addresses the origins of the problem.

- Thanks to the billion-dollar "superfund" program of the US Environmental Protection Agency (EPA),[42] cleanups in the USA, such as in the Hudson River and Puget Sound, where the polluter has even paid most of the costs, have been getting PCB levels down consistently for decades. "The US is going way beyond the Stockholm Convention because they know how toxic PCBs are," comments Paul Jepson at the Zoological Society of London. "All we have done in Europe is ban them and then hope they go away."[43]
- There are many initiatives cleaning up the seas, for example:
  - Friends of the Earth has been urging the UK government since late 2018 "to consider a number of measures to tackle car tire pollution, including a standardized test to measure tire tread abrasion rate and a car tire levy to pay for research into solutions, and to consider mitigation measures."[44]
  - Greenpeace is running a plastics campaign: "From bottles to packaging to microplastics, companies need to take responsibility for what they produce; governments need to legislate for change—and all of us need to change how we think about plastic."[45]
  - A single man began to clean up Mumbai Beach. Now it has become a movement.[46]
  - Launched in 2015, the Global Ghost Gear Initiative brings together governments, private sector corporations, the fishing industry, non-governmental organizations and academia to tackle the problem of lost and abandoned fishing gear (nets, long lines, fish

traps, lobster pots, buoys, etc.).[47]

- The Clean Seas campaign was launched by UN Environment in 2017 with the goal to eliminate major sources of marine litter by 2022.[48]
- Scientists call for plastics to be better designed to encourage recycling. For example, *clear* plastic bottles have a recycling value five times higher than those that have been dyed (because the pigment is hard to remove).[49]
- On his own initiative a young man from the Netherlands developed a global cleaning project: The Ocean Cleanup (www.theoceancleanup.com).[D] Whatever the outcome will be, it is also an encouraging story because it shows how a single (and young) person with a good idea can make a difference.

However, cleaning can be only one part of the solution. It should not encourage the plastic producers to carry on regardless because someone will be cleaning up the mess they make. We need less single-use plastic products, and a new different breed of long-term quality products. The materials should be reusable, recyclable and/or truly biodegradable.

- A main solution to avoid plastics in the ocean is better waste collection and recycling. Opportunities to capture microplastics through enhanced washing-machine filtration systems and improved waste and water sewage treatment processes must also be explored.
- European scientists have urged the car industry to re-think tire materials and their amalgamation, and the EU to decree that tire labelling has to state another quality: longevity.[50] Until then, there is room for some improvement: In autumn 2020, The Tyre Collective, a group of masters students from Imperial College London and the Royal

College of Art, won the James Dyson award for a device that captures tire wear particles right at the wheel. The device is fitted to each wheel and uses electrostatics as well as the air flows around a spinning wheel to collect particles as they are emitted. The prototype collected 60 percent of airborne particles on the test rig.[51]

- What's needed is a radical change in all the industries to produce better, longer-lasting quality products which at the end of their lifespan can be completely dismantled and recycled. *Materials must either be able to enter a biological cycle or continually stay in a technological one.* Nature is the guide: waste is always food/raw material for another type of circulation. Car tires should only shed materials which are biologically degradable. This **Cradle-to-Cradle** approach was already suggested in a 2002 book of the same name.[52]
- An additional angle to fundamentally change consumer waste society is the distribution method. A washing machine bought by a single user might indeed work perfectly—for one month longer than the guarantee is valid. Followed by ten more years with regular costly maintenance. This is a business plan that encourages producers to manufacture short-lived items with inherent fault-clocks ticking. Even better would be a lease model where the producer remains the owner of the device, maintains it and leases it to the consumer. Miraculously, such devices would not need many maintenance visits (since now the producer has to pay for the care plan). Products of unheard brilliance and quality would emerge.

What can I do?

- Avoid, avoid, avoid. (If you want more ideas you can ask

your local bookstore about books dedicated to plastic avoidance.)

- Don't trust plastics labelled "biodegradable". Different types of "bioplastics" (made from recycled material or plant cellulose) are just as toxic as conventional ones. Most of them contain more than 1,000 different chemicals, some of them as many as 20,000.[53]
- Watch David Attenborough's documentary *A Life on Our Planet* (2020). And the video clip about diving in the plastic ocean off the Bali coast: "Bali: Diver films 'horrifying plastic cloud' – BBC News" on YouTube.
- Ask your local water supplier what your degree of plastic contamination is. Ask them what they are doing to keep tomorrow's water clean (hoping that big numbers of regional waterworks begin to put pressure on governments).
- Support petitions and activism for clean water.
- Support towns and cities planning to introduce public drinking water stations, to end the culture of plastic bottles.
- If you like reading small print, avoid the following ingredients when buying cosmetics: PE, PP, PET, Nylon-12, Nylon-6, PUR, AC, ACS, PA, PMMA, PS, PQ.
- Smokers, don't discard filter cigarette butts onto the soil! A single synthetic filter contains enough microfibers to contaminate 500 liters of groundwater.[E,54]
- Watch the documentary *TRASHED*, with Jeremy Irons.[55]

Nano waste

Nanotechnology is manipulation of matter on an atomic and molecular scale. One nanometer (nm) is one millionth of a millimeter. The main material is carbon, the main application so far is carbon nano tubes (CNT) to strengthen plastics in electrotechnics and airplane construction. Nano materials are barely water soluble and they are bound to last for thousands of years. With a worldwide production of 1,000 metric tons of CNT per year, tendency: steeply rising, invasion of the living world has begun.[56]

There is *no* technology to even measure nano particle contamination levels. However, lab studies have shown that the biological behavior of nano particles is somewhat similar to that of microplastics. A Chinese study has revealed that copper oxide nano particles penetrated the membranes of cyanobacteria and damaged their DNA.

It was found that even with their small size nanoparticles dissolved in sea water absorb significant amounts of light energy, which is lost to the microalgae which represent the foundation of maritime food chains—and which produce considerable amounts of the world's oxygen. Nanoparticles also lump together with other particles, thus stealing more light from the photosynthesizing algae.

Scientists in Switzerland tested two algae species (*Chlorella vulgaris* and *Pseudokirchneriella subcapitata*) whose growth among 5.5 milligram of carbon nano tubes was stunted by no less than 75 percent. This is a concentration about 100,000 times higher than nano pollution as of yet, but we should be concerned before it's too late, for once. Also, as of yet, there is absolutely no regulation of production nor emission. Chemist Fabienne Schwab, Swiss Federal Laboratories for Materials Science and Technology: "CNT will probably be even more difficult to remove from the living world than conventional chemicals. Their effects are unclear today, but ideally, we shouldn't let them get out in the first place."[57]

Chemical

Due to the application of vast amounts of **pesticides** (ca. 80 percent of which are herbicides, the rest is insecticides, fungicides, and other specialized substances of chemical warfare), the biggest chemical polluter is agribusiness. There is no need here to repeat the facts and arguments of the current debates about *glyphosate* and *neonicotinoids* which, tragically, experience delay after delay in their phasing out. On the contrary, the dosages are getting higher.[F]

Of all pollinating insects, honeybees are the most important and the most loved. About one-third of what we eat can grow because they pollinated the flowers. And bee populations around the world suffer greatly from insecticides (and other forms of abuse). Humanity will starve without insects. We've already lost three-quarters of them. I couldn't make a better plea for the bees than the 2012 documentary film *More Than Honey*.[58]

The various pesticide groups are not as cleanly target-orientated as they sound. Agribusiness is not about sophisticated surgery but about unleashing weapons of mass destruction. Herbicides don't only kill "herbs" but also ground beetles, sawflies, and wasps, and not just in the sprayed fields but a long way downwind too. They are also responsible for extremely varying sex ratios in certain frog species. And insects poisoned by insecticides are death bait for spiders, birds, and bats. Fungicides have recently been linked to falling bumblebee populations; although on their own they are relatively harmless to the bee family, many interact chemically with insecticides, fatally increasing their toxicity.

And for what? A 2017 UN report has shown that pesticides are *not* necessary to feed the world; even neonicotinoids don't produce consistent benefits.[59] The UN report says that pesticides have "catastrophic impacts on the environment, human health and society as a whole"—including an estimated 200,000 human deaths a year from acute poisoning. The UN report blames manufacturers for "systematic denial of harms" and "aggressive, unethical marketing tactics."[60] But UN reports have little power; from 2000 to 2018, the global chemical industry doubled in volume.[61]

The poisons are not the only problem. Only a fraction of the lavishly applied nitrate and also phosphate fertilizers stays on the field; much is washed out and causes chaos in the riverine and sea water chemistry (*eutrophication*); algal blooms are only the tip of an iceberg that amounts to ecological collapse of local marine spaces—the so-called (and expanding) "dead zones" in the ocean.[62] This is a silly waste too, as phosphorus is a mineral that is essential to life but scarce: nobody knows how many decades' worth of phosphate are left in the Earth's crust.[G]

And worst of all, the predominant nitrate fertilizers cause serious contamination of groundwater and surface water—a major problem in agricultural landscapes (made worse by the slurry from industrial livestock farming). In Germany, groundwater frequently exceeds the EU limit of 50mg nitrate per liter[63] (for which it is standing trial at the European Court of Justice[64]). The Swiss limit is 25mg/l, the US limit is 10mg/l.[65]

Since the millennium, New Zealand is catching up fast with contaminating its groundwater; with farmers using the highest amounts of nitrogen fertilizer per acre in the world: three times as much as in Britain, and even twice as much as in China.[H] Over 15 years, New Zealand has lost its healthy drinking water, and in 2018 its "pure nature" image too. The land of sheep keepers has become the land of cattle breeders because China buys milk powder and fresh beef in vast amounts. 4.7 million people now share the land with 6.6 million cows, each of which produces up to 50kg of feces a day. The alarming nitrate values of soils and feces-rich lakes now force many New Zealanders to drink from (plastic) bottles only.[66]

Oblivious of the mistakes being made in New Zealand, Ireland followed suit. Its government's ten-year growth agenda from 2010 set out to increase Ireland's food production by a third. New subsidies were created to push up beef by 20 percent, dairy and pork by 50 percent, and make Ireland an international food player. Fully green-washed of course, "sustainable" growth selling products with an overall "positive branding of Ireland as green and clean."[67] Ignoring the fact that livestock at an industrial level is a contradiction in itself to responsible ecosystem care. A decade on, nitrate and phosphorus pollute the rivers and water pathways, and antibiotics from intensive farming begin to infiltrate the groundwater. Glyphosate runoff from equally destructively managed cereals and energy crops adds to the contamination of waterways. The number of pristine rivers in Ireland has fallen from 575 in the 1980s to an all-time low of just twenty in 2020. Furthermore, the agenda's promise to encourage the planting of "more broadleaf varieties to improve biodiversity and leisure benefits" manifested as the opposite: far-reaching Sitka spruce plantations, ecological dead zones which destroy habitat for endangered species such as hen harriers and curlews. But don't blame the farmers, as usual it is the governmental system of subsidies that sets the direction.[68]

The over-poisoned killing fields of industrial farming not only "commit insecticide" but destroy everything in their wake. That includes the microorganisms in the soil. The healthy humus disappears with them (releasing gigantic amounts of carbon in the process), and soil erosion further degenerates the ground from which we live. The third disaster concerning the soil in industrial farming is ground compression caused by the heavy machinery. Compacted ground has much less oxygen and develops an altered chemistry. None of this is benevolent to soil organisms; biodiversity and abundance are decreasing. Furthermore, strange decisions in (or omission of) crop rotation and the failure of planting ground cover between crops (particularly bad with maize/corn) cause far more soil erosion than necessary. All in all, due to these failures of industrial agribusiness, **the world had lost a third of its soils by 2014**; we are now approaching half. *If the current rate of total unsustainability continues, the last fertile top soil will disappear in sixty years.*[69] The destruction of soil is as threatening as the climate emergency. "We are losing 30 soccer fields of soil every minute, mostly due to intensive farming," says an expert of the International Federation of Organic Agriculture Movements.[70] It takes thousands of years for healthy soils to form. The global food system is broken.[71]

Furthermore, pharmaceutical drugs enter soils and the water cycle everywhere. Some 150 active ingredients have been found so far.[I] But what do frogs say about the glut of painkillers, statins, beta blockers, anti-diabetics, antidepressants and other psychoactive medication? Their tadpoles often develop deformed bone structures. Amphibia are particularly sensitive to *endocrine disruptors*. This group includes PCB, DDT, bisphenol A, phthalate and many others. In starfish, both freshwater and sea snails and fish, amphibia, reptiles, birds, and mammals they lead to infertility, deformations (especially of the sexual organs), and gender changes (masculinization, feminization).[72]

A special case is antibiotics.[73] Europe and the USA alone use about 20,000 metric tons every year, half in human medicine, the other half to keep the crowded beasts in industrial livestock farming alive (still, 91 percent of the pigs in livestock farms suffer from joint inflammations (synovial bursa), 44 percent of them to a medium to high degree; almost 14 million pigs die every year in German meat farms alone).[74] Worldwide, nearly three-quarters of antibiotics are used on livestock.[75] A lot of this enters the soil and water cycle with the slurry. Even in "clean" Switzerland, not exactly a meat export country, the antibiotics contamination reached 1kg per hectare (2.5 acres) already in 2001. This is enough to kill many soil bacteria—or breed resistant strains.[76]

Antibiotics that make their way down the watercourses also contaminate estuaries, lagoons, and ultimately, the sea. Samples from the blowholes, gastric fluid and feces of 171 bottlenose dolphins from the Indian River Lagoon in Florida showed that by 2015, 88 percent contained antibiotics-resistant bacteria.

The antibiotics came from agricultural runoff but mostly from sewage discharge. Once in the lagoon, the antibiotics create selective pressure on the normal bacteria which are reduced by the antibiotics. Resistant bacteria remain and proliferate, creating populations of resistant pathogens that maritime organisms are exposed to.[77]

What is being done?

A first understanding is beginning to dawn that the entire unsustainable behemoth of industrial agriculture has to change, away from heavy machinery and mineral oil (artificial fertilizers and pesticides). But how long will it take? The alternative is ready and waiting: **localized, organic farming with little or no tilling**. And techniques like biological pest controls, polyculture (growing multiple types of plants), permaculture, crop rotation. Also, three-quarters of human food is actually produced by small farmers anyway; Big Agro mostly produces bulk grain for unhealthy carbohydrates, maize for poisonous sugar products or for dubious biogas plants, and, most of all, cattle fodder.

What can I do?

- Support organic farming as much as you can.
- Never take antibiotics lightly. They also reduce the biodiversity of your healthy gut flora (microbiome).
- Buzz with the bees.

Genetics

Monsanto and the huge areas of genetically modified (GMO) crops are another global experiment to see how unnatural DNA will survive in, or possibly interfere with Earth's ecosystems. The original promise of the industry was twofold: genetic modification would make crops immune to certain pests and hence reduce the need for pesticides, and, first and foremost, the world's growing population supposedly *needs* this new technology in order to be fed sufficiently. About three decades later, this bubble is bursting for good. An extensive examination by *The New York Times* (using data from the United Nations and the National Academy of Sciences) shows that *genetic modification has not accelerated increases in crop yields*[J] *or led to an overall reduction in the use of chemical pesticides*. On the contrary, pesticide sales have strongly increased.

Genetic modification sparked an early discussion about ethical limits—at least in regards to modifying human DNA—and as a result, various legal frameworks to keep modern biotechnology in bounds were established. In contrast, with the new GenDrive technology and its associated "gene-editing" system CRISPR/Cas9, humanity can soon accidentally or intentionally wipe out whole populations, even entire species, and as of yet there is no legal framework to control, direct or limit this highly controversial "new genetic extinction technology."[78] The ethical debate is still wanting.

It sounds advantageous to eradicate an invasive species to protect a native ecosystem, or "weeds" to give a hand to farmers and thereby reduce pesticide use, or even erase those two or three mosquito species which are the main carriers of malaria, zika, and dengue fever. After all, the mosquito *Aedes aegypti*, by spreading malaria, is the only creature responsible for the deaths of more humans than humans themselves. But once it is tried, tested and proven, are we able to confine it to mosquitos? This same technology can serve as a means of biological warfare and, for example, destroy the harvests of enemy countries. Indeed, the US military is highly investing in GenDrive technology.

At the UN Convention on Biological Diversity (CBD) in December 2016 in Cancun, more than 160 NGOs demanded an urgent, *global moratorium* on the development and release of GenDrive technology, because it poses serious and potentially irreversible threats to biodiversity, as well as national sovereignty, peace and food security. Among the NGOs were Friends of the Earth International, the International Union of Food Workers (representing over 10 million workers in 127 countries), and the European Network of Scientists for Social and Environmental Responsibility. Following this call for a global moratorium, the CBD, as the key body for GenDrive governance, began to search for scientists and experts as advisors in a task force.[79]

Ten months after Cancun it came to light that the Bill and Melinda Gates Foundation had paid $1.6 million dollar to a private PR firm to work "behind the scenes to stack key UN advisory processes with GenDrive-friendly scientists", and thus "to counteract proposed regulations and to resist calls by scientists and conservationists for an international moratorium." It was also revealed that the US military (the Defense Advanced Research Projects Agency, DARPA) is now the top funder and influencer of the accelerating GenDrive development—with a budget of $100m.[80]

Radioactivity

The "peaceful" use of nuclear energy has not been looking that peaceful anymore since the disasters at Chernobyl in April 1986, and Fukushima in March 2011. If anyone still thinks nuclear is a "clean" energy they should just google "Chernobyl mutation" images (or go to the source: Chernobyl Guide: chernobylguide.com/chernobyl_mutations/)—and realize that such deformed animals and children are still being born in the region, decades later.

Both the Chernobyl and the Fukushima disasters spread their highly radioactive load around the entire planet. The worst-case nuclear disaster occurred in March 2011 in the power plant at Fukushima—still leaking incessantly many years later—contaminating rural fields, villages, soils, tap water, sea water and marine life in the northern Pacific. Ocean currents carried and carry the radiation eastward, much to the detriment of fisheries along the North American west coast. 120,000 Japanese remain displaced and Japanese taxpayers face a bill that will run to hundreds of billions of dollars.

Certain unstable atoms produce ionizing radiation which consists of energetic subatomic particles (ions or atoms moving at incredible high speeds) and of electromagnetic waves on the high-energy end of the electromagnetic spectrum. In organisms, even at low doses they can cause DNA damage, mutation of reproductive cells, and cancer (BfS 2018). With a radius of 30km around the power plant of Chernobyl the "zone of alienation" has largely reverted to forest, with abundant wildlife such as deer and wolves. An abandoned car deck offering winter shelter to a pack of wolves appears somewhat apocalyptic, but overall the "death zone" *looks* surprisingly green and lush. However, all species suffer from various mutations large and small. Also, due to global heating, forest fires are beginning to spread radioactive isotopes into the atmosphere once again; serious wildfires occurred in spring 2020.[81]

According to the German BfS (Federal Office for Radiation Protection), wild mushrooms and game in Germany are still contaminated with caesium-137 from Chernobyl. Although the substance has now reached its 30-year half-life, the recommended limit of 600 Becquerel per kg can be exceeded, especially for wild boar meat. However, radiation is strongly region-specific, as some areas were more severely affected than others at the time. An extensive meal of forest mushrooms with above-average contamination corresponds to the radiation exposure of a flight from Frankfurt to Gran Canaria (although when flying, the radiation is evenly distributed over the entire body, while contaminated food primarily affects the digestive system).[82]

Nowadays, in the discussion about phasing out fossil fuels, some people have begun to praise nuclear power again, for its low carbon emissions and as "clean" and climate-friendly energy. They are anything but. Certainly nuclear power stations cause smaller carbon emissions than coal plants, but just the vast amounts of concrete for their construction, for one, render them anything but climate-friendly. Their footprint when they are up and running isn't fabulous either: nuclear energy based on high-grade uranium ore produces average emissions of 60 grams of CO_2 per kilowatt hour of electricity, compared with 10-20g per kWh for wind energy and 500-600g per kWh for gas. As global uranium reserves dwindle, more and more diesel fuel is needed to mine and mill the uranium, hence a nuclear power station running on low-grade uranium ore has been calculated to emit ca. 131g per kWh, which is seven to 13 times as much as wind energy.[83]

Furthermore, nuclear power doesn't even make sense on the purely economic level: their true overall cost will never amortize. In the USA, about 80,000 metric tons of used spent nuclear fuel are stored at more than 75 sites in 35 states around the country. After more than half a century of effort nobody in the USA, or anywhere in the world, has developed a successful plan for a final solution for the management of nuclear waste. The US Department of Energy spends about half a billion dollars every year on the facilities where the nuclear waste "temporarily" remains.[84]

In the UK, the 2012/13 budget of the Department of Energy & Climate Change (DECC) allocates £2.5 billion a year to nuclear decommissioning and waste handling. £1.6 billion of this is spent on Sellafield (Cumbria) alone, home to the radioactive remains of nuclear weapons and energy programs dating more than half a century. The complete decommissioning of Sellafield will take over a century and will cost an estimated £67 billion.[85] And it's not just £2.5 billion a year for the next hundred years—the half-life of plutonium is 24,000 years.

Sonar and noise pollution

Similar to sound in the air, a sound wave underwater consists of alternating compressions and rarefactions. Organisms detect these compressions as changes in pressure on their skin, and in their ears (if they have any) while deeper sounds can be felt throughout the body. Sound propagates far underwater; the frequencies best suited to water are those between 10Hz and 1MHz. Sound speed in water is more than four times higher than in air. Both the water surface and the sea floor are reflecting (and scattering) boundaries. Fish and most marine creatures have sensitive hearing. Dolphins and other toothed whales have very acute hearing sensitivity, especially in the frequency range 5 to 50kHz.[86]

Whales are known for their far-traveling calls and songs, communicating to fellow members of the pod. But scientists have now revealed a unique, intimate form of communication between humpback mothers and their calves, as well as a silent method to initiate suckling. Because their breeding waters (some 5,000 miles away from their Arctic or Antarctic homes) are beset with killer whales preying on stray calves, and also to avoid males who are still looking for mates, humpback mothers and calves "whisper" to each other.[87]

Noise from engines and large propellers of commercial or naval shipping, which has increased dramatically over the years, harms whales and dolphins. Group hunting by dolphins or orcas is also impaired because the animals' acoustic signals drown in the background noise. Seals are being temporarily deafened by noisy shipping lanes.[K] What's more, the whiskers of seals are tactile sensors which help the animal to "see", dive and hunt even in the dark. They vibrate at frequencies of 100-300Hz,[88] a frequency band nowadays being highly drowned by engine noise from ships.

Worse than ship engines is seismic exploration in the manic hunt for new mineral oil or gas fields. Pneumatic guns (air guns) and explosives create low frequency sounds (beneath 100Hz) to provide single pulses or continuous sweeps of energy which generate seismic waves to probe deep into the seabed. Air guns with up to 250 decibels are about one thousand times louder than a ship engine, and tear everything in their vicinity to tatters. In the Arctic, seismic exploration for oil and gas is nonstop. And if a site tests positive the arrival of the oil rig will bring permanent noise pollution to the marine populations.

The worst is the loud military sonar used to detect submarines. It amounts to nothing less than torture for whales and dolphins who are highly sensitive to sonar. A single sonar device can be heard in an area of 300,000 square miles (twice the size of California).[89] Sonar has been identified as the main reason for their unnatural mass strandings. Beaked whales, the most common casualty of the mass stranding events (MSE), have been shown to flee from these sonic activities. And even blue whales, the largest animals on Earth, and whose population has plummeted by 95 percent in the last century, are known to abandon feeding and swim rapidly away from sonar noise. Mass strandings with multiple species of whale and dolphin have soared since the introduction of military sonar in the 1950s.[L,90]

Despite all this, in 2015, the US Navy ordered 136,000 sonar buoys to be spread around key positions in the world's oceans. Permanent surveillance of the seas shall facilitate the swift detection of "enemy" submarines.[91] Surprised by sonar, whales panic and rise from deeper waters too fast and the change in pressure causes internal injuries. Whales are only the visible fraction; noise pollution affects all marine creatures negatively.

Following a study by Fauna and Flora International (FFI),[92] Sir David Attenborough urged governments in spring 2020 to ban deep sea mining. Sonar and machinery on the seabed can jeopardize the ocean's life support systems by causing significant loss of biodiversity, disruption of the ocean's biological pump, and the loss of microorganisms important for storing carbon.[93]

What is being done?

In June 2018, the Spanish government decreed a reserve for whale migration thanks to Teresa Ribera, Minister for the Ecological Transition of Spain (Ministra para la Transición Ecológica),[94] between the Balearic Islands and the mainland. The protected area covers 17,909 square miles (46,385 square kilometers). The timeline is of note: the UNEP Convention for the Protection of the Marine Environment stems from 1978, Spain ratifies it in 1999 and acts on it in 2018.[95] The species Spain hopes to protect from sonar and pneumatic devices, detonations and oil drilling are fin whales, sperm whales, pilot whales, Cuvier's beaked whales, bottlenose dolphins, striped dolphins, common dolphins and loggerhead turtles.

In June 2018, the Canadian government closed a strip of water along the southwestern coast of Vancouver Island to commercial and even recreational fishing. The protected area, about 45 nautical miles long and 6 nautical miles wide, is intended to help the struggling orca population known as the "southern resident killer whales", to find more of their main food source, Chinook salmon, and eat in peace. However, the tranquility is frequently interrupted by fire exercises by the Canadian navy and the US Coast Guard.[M]

And of course there are other marine reserves, for example in New Zealand and Costa Rica.

What can I do?

- Support marine creatures, e.g. check out Greenpeace projects or whales.org.
- Avoid yachting and jet-skiing.

Light pollution

As the natural rhythms of darkness and light govern most of life on Earth, the light pollution from cities confuses or even impairs the mating and other behavioral patterns of many species, particularly insects and birds.

Every little helps: close your curtains at night.

Microwaves

Since life appeared on planet Earth, the natural radiofrequency environment comprised of broadband radio noise from space (galactic noise), lightning (atmospheric noise), and a small amount of radio emissions from the sun (solar noise). Plants and animals have evolved in harmony with this *weak* natural radiofrequency background and use the periodic nature of its fluctuations to regulate some of their metabolistic functions. But in the 1920s, human-made radiofrequencies became a worldwide phenomenon and their intensity has been constantly increasing ever since. From the perspective of evolution, this change is sudden and dramatic, living beings have had no chance to adapt to these new and challenging conditions of electromagnetic field (EMF) pollution.

Depending on intensity and exposure time, colonies of various types of bacteria can be restrained, decimated or even wiped out by certain high frequencies;[96] others, for example lactic acid bacteria, may even flourish.[97] Electromagnetic signals with high frequencies can also trigger (or eliminate) antibiotics resistance in bacteria[98] (a growing problem, particularly in high-tech hospitals). One begins to wonder: how does the ever-increasing electromagnetic pollution change the bacteria in our gut, and the other microbes which are essential for our health? And what about our own cells, and our own DNA? It becomes apparent that blindly interfering with these processes of the living world is a very dangerous game.

Electromagnetic pollution disrupts vital cues in the life

processes of marine creatures. So far this has been shown for American lobster and little skate.[99] Anthropogenic sources of marine electromagnetic fields include ships, bridges, and subsea cables.[N]

A few examples from the plant world. Soybean seedlings show reduced growth when exposed to cell phone (4G) microwave radiation (900MHz).[100] Electromagnetic radiation induces stress in Scots pine trees (*Pinus sylvestris*): it diminishes the creation of lignin (key structural component of cell walls) and accelerates resin production, resulting in premature aging.[101] The trembling aspen (*Populus tremuloides*) decline which has been increasing across North America since the 1950s has been linked to radiowave pollution. Test seedlings which have been shielded from the artificial "background noise" which is omnipresent today (from 1MHz to 3GHz) developed 60 percent more leaf area than the exposed trees.[102]

It has been shown that birds lose their sense of orientation when they are exposed to artificial electromagnetic disturbances. Frequencies in the range of AM radio—*a hundred to a thousand times below the guidelines adopted by the World Health Organization to protect human health*—proved enough to disable the magnetic compass of robins.[103] Weak radiofrequency magnetic fields (190 microtesla at 1.4MHz) disrupted the orientation ability of garden warblers (*Sylvia borin*).[104] A Spanish study showed that microwaves are significantly interfering with the reproduction of white stork (*Ciconia ciconia*).[O,105]

Due to their size, insects are affected disproportionately. Radio frequency radiation from cell phones influences behavior and physiology of honeybees (*Apis mellifera*). An Indian study[106] from 2001 showed an initial reduced motor activity of the worker bees on the comb when exposed, followed by restlessness and en masse migration. Metabolic changes comprised of a rise in concentration of biomolecules including proteins, carbohydrates and lipids. It is no surprise bees are sensitive to electromagnetic phenomena because, apart from their magnetoreceptor in their abdomen, their integument (outer shell) has semiconductor functions; and the antennae of flying insects—sensors for smell, taste, humidity and temperature—are part of the bioelectrical nervous system (the changes of the electrical potentials of the antennae to different odor can even be measured).[107]

A 2018 Swiss study confirmed that EMF pollution is harmful to insects. Frequencies above 6GHz lead to changes in insect behavior, body shapes and functions. The amount of absorbed energy by the insects can increase up to 370 percent—which will fry them in mid-air.[108] Global heating, insect heating. And birds and animals too. We are set to worsen the fever of an already imperiled biosphere.

What is being done?

Despite their obvious similarities, and word going round that using a cell phone "fries your brain", we don't really want to distance ourselves from our favorite toy. Nor does the industry: selling smartphones worldwide is a multi-hundred-billion dollar market. The electronic sector has a strong lobby, you can tell by the fact that radiation limits are not being reviewed, and critical scientific studies hardly ever find an echo in mainstream media. From the outset, cell phones had an easy start: they were allowed, remarkably, to enter the US market and many others around the world without *any* government safety testing.[P,109]

It's not that there was no science on the matter. Studies indicating malevolent effects of radio frequency (RF) radiation have been around for decades. For example, in 2003, the Swedish neurosurgeon Leif Salford and his team found that weak pulsed microwaves give rise to a significant leakage across the blood-brain barrier, as well as "highly significant evidence for neuronal damage in the cortex, hippocampus, and basal ganglia in the brains of exposed rats."[110]

In 2011, the World Health Organization (WHO) reversed

its previous verdict (from 1927!) and declared radio frequency (RF) radiation to be dangerous, namely a class 2B "possible carcinogen", in the same category as lead and nickel.[111]

A US metastudy in 2017 looked into the effects of cell phone emitted radiations on the "orofacial structures" (oral cavity and lower face) of humans. It found that radiation has an adverse effect on oral mucosal cells, facial nerves, salivary glands and causes changes in salivary flow rate.[112] However, scientific studies of the effects of radio waves can hardly find control groups with no exposition because the whole surface of our planet is continuously covered with radiation from a whole number of satellites.

And in March 2018, when a landmark US government study of the health effects of cell phone radiation (900-1900MHz) was released, "not one major news organisation in the US or Europe reported this scientific news." Until finally *The Guardian* had the courage to publish it four months later ("The inconvenient truth about cancer and mobile phones"). The study, ranking among the largest conducted on cell phone radiation, concludes that there is "clear evidence" that radiation from cell phones can cause heart tissue cancer, and evidence of cancer in the brain and adrenal glands.[113]

Mobile telecommunication will be with us for some time, one way or another. But it would be possible to develop technologies that are less detrimental to organisms (e.g. glass fibre cables). Instead, 5G is being pushed. Billions of dollars leave no space for irritating questions about negative effects on humans, bees and birds, and other living things. But they are to be expected, particularly since the high frequency signals of 5G emitter stations will blow out vast amounts of electrical energy. High frequency does not reach far, which necessitates (as a downside for us) a much denser network of masts and stations than the previous mobile communication technologies.[Q]

What can I do?

- Never trust a hype, particularly if those who tell you to be gleeful are the ones to profit from it big time.
- Use all wireless phones sparingly. To protect your own head, whenever you can use the loudspeaker, use it. Radiation levels decrease to the square of the distance, i.e. doubling the distance from your ear, say, two inches instead of one inch, results in only a *quarter* of the radiation.
- If you consider a headset, prefer wire to Bluetooth.
- Turn your WiFi modem/router off, at least at night. The insects and birds outside your building will be a bit better off too. For Christmas, you could buy your neighbors a little pocket guide about electrosmog.
- If you feel uneasy about the rollout of 5G you can visit the large global resistance movement which organizes an international appeal, "Stop 5G on Earth and in Space": https://www.5gspaceappeal.org/.

Health

The Earth needs healthy Earth citizens, so let us not overlook our own well-being. It is our Gaian duty to care for ourselves too.

Biodiversity, again

Of all the biota in the biosphere, 13 percent of the living weight is bacteria (the overall global biomass of bacteria is about one thousand two hundred times bigger than that of humans).[114] The arrangement of the human body is quite similar: about 4-6 pounds (2-3kg) of the weight of a grown-up is bacteria. They are everywhere. Covering our entire skin, protecting it from microscopic invaders the way forest fungi protect "their" trees. Bacteria are present in all our organs, even the brain, and especially the colon. Billions of little helpers without whom we would not be alive. We have about one trillion body cells, but ten times more bacteria. In biology their grand total is referred to as the human *microbiome*.

Much to the surprise and frustration of the scientists involved, the Human Genome Project at its completion in 2003 found that the human body, the most complex and amazing organism in the galaxy, has only just over 22,000 genes in its cells, the same amount as in other mammals. How could the human have the same amount as a house mouse? On second glance, the house mouse even has 3 percent more protein-coding genes than we do.[115] So *where is* the blueprint for our body and its functions?

Astonishingly, it is the DNA of our bacteria which contains the genetic information that makes us what we are. Together, all of our bacteria carry a hundred times more DNA than our human cells. In terms of cell count, we are ten percent human. In terms of DNA, we are one percent human.[116]

Our colon microbiome does most of the digestion for us. If it is in good shape it has a tremendous positive influence on our immune system, and it also impacts the other organs, hormone activity and thus some of our moods. It also has feedbacks on our DNA and protein generation.[R]

And diversity is key. The greater the variety of species in and on us, the better our digestion and our immune system will work. At birth we get prepared for life by receiving a treble imprint with the mother's microbiome: in the birth canal, via skin-to-skin contact straight after birth, and via the breast milk. This is the legacy of the maternal line, reaching back thousands of years! Caesarians and formula milk cannot deliver this.[S]

The final step for a healthy microbiome for life is to be found in the childhood environment. A study in 19 EU countries found that children living on a farmyard have significantly less risk of developing asthma or hay fever than city children, simply because of the higher diversity of microorganisms. Bedbugs, no joke, protect from asthma.[T] It is no surprise then, that babies and toddlers constantly follow a mysterious urge to run around and put everything into their mouth or lick it. They are imprinting themselves with as much of bacterial diversity as possible.

So forget their bad reputation! Only a tiny minority of bacterial species are harmful to humans. Most of them we could not live without. Therefore the use of antibiotics (from Greek, "against life") is a measure of vast collateral damage in the organism, with deep side effects and chain reactions, and of widely unknown long-term consequences—although some of which are beginning to come to light these days.

The great detox

The situation we are in today is that our bodies, like all organisms on Earth, are exposed to a bombardment of over *30,000 artificial substances* which humanity has created since the beginning of the Industrial Age. Only about 1.8 percent of them are *not* toxic for us. And because they are brand-new in terms of the evolutionary time-scale, the ancient wisdom of our bodies doesn't quite know how to deal with them. Our immune systems are over-burdened with the constant influx of microscopic invasive aliens. Under this permanent siege by "environmental" toxins, our immune systems fail in other places and we develop food intolerances, and allergies to previously harmless natural substances that never posed a problem to our species but now trigger partial collapses in our immune systems. Like all our sisters and brothers in the animal and plant kingdoms, we are fired at by the whole spectrum of pollutants, plus the—mostly unknown—biochemical interactions between them.

In the media coverage of the global plastic tsunami so far, we often read or hear that "little is known" about the actual health effects of the many different microplastics and chemical pollutants entering the organism. At best we hear that certain substances "accumulate in fat tissue". Sounds rather peaceful,

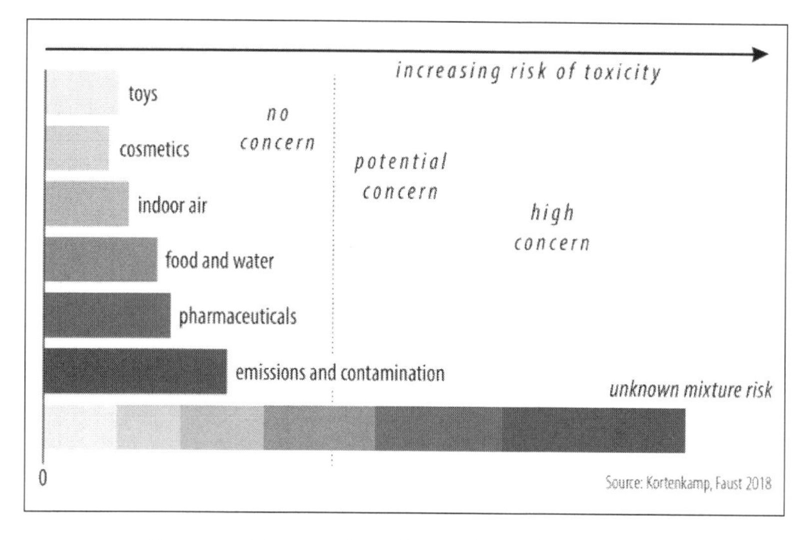

Figure 7: The combination effect of pollutants

All life forms today are exposed to tens of thousands of chemicals from many different sources. Despite growing scientific evidence for enhanced toxicity of cumulative chemical mixtures, regulations do not adequately reflect the dangerous combination effects.

perhaps our bodies can tolerate the toxins just hanging out there in our fat depots? And the levels found in urine and stool samples[117] are good news too, aren't they? It seems the stuff is leaving our bodies. But there are three major points of deep concern:

- We don't know the proportion of how much our body can expel and how much it cannot.
- We don't know how much energy it costs our body to activate the spectrum of immune responses first and eventually process the toxins through the lymph system, the liver, the kidneys and the colon. It certainly is highly demanding on the immune system.
- Furthermore, certain toxins get processed in the liver, but in the colon—after seriously irritating some of our

allied bacterial tribes—pass the colon-blood barrier into the portal vein which carries them back to the liver! They can get trapped forever in the so-called *entero-hepatic circulation*.

To say that nothing is known about microplastics and other nano pollutants in the body is not honest. For a while now, some medical laboratories have been working with mass spectrometry, a method for measuring the mass of molecules that actually comes from physics. For the investigation of blood plasma, they offer to *analyze the mitochondrial DNA from leukocytes*. To recap: leukocytes[U] are the white blood cells, an important part of our immune system. The mitochondria are the small "power stations" in every cell that supply them, and thus us, with energy (see "Teaming up" in Chapter 4).

Heavy metals such as lead, mercury and cadmium are tested, as well as chemicals[V] such as methyl acrylate, phthalates and dichlorophenol—the list of pollutants in our bodies does not surprise us anymore, but the wake-up shock follows regarding their location: *single molecules of a vast range of toxins can be found docked on to the base pairs of the DNA*. They should actually be completely free: when the cell needs new proteins (basically, all the time), it sends messenger substances to the DNA to retrieve the blueprints. To do this, the messengers dock to the open ends of the DNA base pairs (as when we plug a USB stick into a computer to copy and take information with us). If the contact points are now occupied, this leads to serious problems.

I interviewed a specialist for preventive medicine, Dr. Frank Borower, about the importance of mitochondrial DNA:

"DNA must be read. Some genes are blocked anyway (that's epigenetics), but nowadays chemicals interfere too. The mitochondrial DNA is easier to access and much more vulnerable to such intruders. However, *the mitochondria are very important because it is here that things like cell growth and cell division are decided*. The mitochondrial DNA also initiates *apoptosis*, which is the 'suicide program' of a cell, so that at the end of its life it shuts down in a controlled manner, switches off and disposes of itself, through an ordered implosion, instead of simply ripping out its cell contents and thus unnecessarily burdening the immune system or even triggering autoimmune reactions. Self-degradation of cells takes place constantly, balancing the renewal by cell division.

This cell degradation no longer works when the mitochondria no longer function properly. Then the cell can no longer adequately bid farewell to the tissue community. With weakened mitochondria, a dying cell lacks the energy for proper self-dismantling. Without apoptosis, the cell basically becomes 'immortal'. This is how a tumor cell develops, so this is also mitochondrially controlled. Therefore, many cancers are certainly 'environmental' diseases."[118]

The mass spectrometry results are doubly dramatic: not only do we learn that the DNA is blocked at different sites, but they also show that the immune system, which already has its hands full, is additionally weakened from the inside out. Dr. Borower talks about the broadband attack caused by harmful substances:

"Our body has to work hard with its systems to cope, especially the detoxification system, which is misappropriated for this purpose. The body's own toxins accrue constantly anyway, that's how our detoxification system has developed in evolutionary biology. But now it is being misused for a whole number of other substances. Nevertheless, the body's own toxins, which also have to be disposed of, still remain. The immune system is often massively irritated when it begins to deal with these environmental toxins as if they were normal pathogens. Because pathogens usually divide and multiply, and can be fought by limiting and containing their growth. But harmful substances behave differently, they do not multiply, and when the immune system starts, incorrectly, to try to solve the problem immunitarily, it does not work. It just consumes an enormous amount of energy, a lot of ATP [the energy 'currency' of the body, see Chapter 3]. This misdirected immune response is probably also a cause of the many symptoms of exhaustion today."

This fundamental weakening of the immune system due to impaired DNA accessibility is a heavy price that not only humans pay, **but all living creatures**. So many tree species succumb to "pathogens", and vast animal populations disappear faster than we can determine why. Global DNA micro-pollution might play a bigger part in the sixth mass extinction than we know.

What can I do?

- Live a life as healthy as possible: good food, good air, good water. And no stress.
- Honor your colon. Be grateful for your microbiome. It keeps you alive.
- Let us find ways together of healing our fellow animals and plants too. Every organism has the right to a healthy life!

Footnotes

A. Plastic trash received by Malaysia alone, in 2018:[119]
  - from the UK (January-August): 88,000 metric tons
  - from the USA (January-June): 150,000 metric tons

B. In the Xingu River, a major tributary of the Amazon. More than a quarter of plastic particles in freshwater fish were polyethylene (a material used in fishing gear), others were identified as PVC, polyamide, polypropylene, rayon and other polymers used to produce bags, bottles, food packaging, etc.[120]

C. Recommended reading about the orca population in the Pacific Northwest: Jim Robbins 2018. Orcas of the Pacific Northwest Are Starving and Disappearing. *New York Times* online, July 9, 2018. https://www.nytimes.com/2018/07/09/science/orcas-whales-endangered.html

D. The Ocean Cleanup https://www.theoceancleanup.com was founded in 2013 by the Dutch inventor Boyan Slat when he was just 18 years old. It raised over 2 million US$ with the help of a crowdfunding campaign in 2014, and has been mainly funded by donations and sponsors since. After a team of 70 scientists and engineers spent five years testing models and prototypes, the first test ship took to the sea in September 2018. The setbacks which occurred had been anticipated. Some critics doubt whether a fleet of plastic-gathering ships might do more environmental damage than good (like possibly harming sea turtles and tuna as "by-catch"), but should we wait until the turtles die of the plastic? Surely it is better to catch it while we can, before it disintegrates into micro- and even nanoplastics.[121]

E. Unfortunately, smoking damages the ecosphere considerably and consumes enormous resources. The climate damage caused by global tobacco consumption is as great as that caused by entire countries such as Israel or Peru. A smoker who smokes 20 cigarettes a day for more than 50 years needs 1.4 million liters of water (which makes the smoker's water consumption ten times higher than that of even a meat eater), a cultivation area of 34,000 square feet (3,200 square meters) and 1.3 metric tons of mineral oil. Thus, the smoking activity of one person causes carbon dioxide emission totaling 5.1 metric tons.

A study by Imperial College (2018, for WHO) currently estimates that six trillion cigarettes are produced each year and smoked by about one billion people. It would make much more sense to use the 9.9 million acres (four million hectares) of tobacco cultivation area in other ways. Prof. Nikolaos Voulvoulis: "It would be much more sustainable to replace tobacco cultivation with the cultivation of other products. That would be better for people's health, but also for the economy there. Given the scarcity of resources, it makes much more sense to grow higher quality plants—that would be good for the local economy, but also for the world economy."[122] And the ecology, as plants which are *not* burned, sequester carbon. The tobacco fields could be reforested.

F. 75,000 metric tons of *active* pesticide ingredients were sold in 2014, according to EU figures. Never assume farmers today poison their estates only once or twice a year—orchards, for example, might get sprayed 25 times (in words: twenty-five) every year between January and August. There is no hoverfly or bumblebee that could survive that. And hence no birds come to the country anymore because there is no food for them. Birds such as tits once used to be a biological check on insect populations; fruit growers used to put up nesting boxes in their trees.[123] Now the birds are disappearing from the landscapes. In France, for example, the dramatic fall in farmland birds, believed to be due to neonicotinoids, parallels the continuing rise in the profits of agrichemical companies.[124]

G. Phosphorus is one of the essential plant nutrients and also used to sustain the human population: surface-mined from phosphate rock it is used to produce phosphate fertilizers. And a lot of it. In 2015, 223 million metric tons of phosphate were mined globally, a number growing by some 3 percent each year. But nobody knows how much accessible phosphate rock is left in the Earth's crust; will it last for centuries, as some say, or just for 20 years? It really should be recycled from slaughter waste and slurry as it used to be, but since a few years now these materials are too contaminated with hormones, pharmaceutica, and heavy metals to go back onto the cropfields(!).[125] Experts are trying to develop new technologies to recycle phosphorus.

H. More information on fertilizers: https://ourworldindata.org/fertilizers
I. Usually amounting to 0.1 to 1 microgram per liter (DVGW 2015), authorities consider them not a threat to human health but try to keep them down (they say).[126]
J. The United States and Canada have gained no discernible advantage in GMO yields (food per acre) when measured against Western Europe where GMO was widely rejected from the start. Overall, GMO has not created "more food for the world".[127]
K. A UK study investigated underwater noise levels in Moray Firth on the north-east coast of Scotland, and found that the noise from shipping was loud enough to cause temporary hearing loss in the seals studied. Dr. Esther Jones, an ecologist from the University of St. Andrews, compares the situation to noise pollution of inner cities: "Urbanisation of the marine environment is inevitably going to continue, so chronic ocean noise should be incorporated explicitly into marine spatial planning and management plans for existing marine protected areas."[128]
L. A study of Cuvier's beaked whales off the coast of Southern California showed that in response to a simulated military sonar signal between 3km and 10km away, the whales initially stopped feeding and swimming, then swam rapidly away from the noise. They also stopped feeding for a whole 6-7 hours, which is unusual. A study of a blue whale showed that, disturbed by the sonar, the animal missed out on about a day's worth of food. To sustain their extreme body size, they have to continuously dive and filter-feed on krill throughout the day. Sonar-induced disruption of feeding could have significant impacts on individual whale fitness and the health of populations.

The US Navy part-funded these studies but said the findings only showed behavioral responses to sonar, not actual harm. Nevertheless, a US Navy spokesman said in 2013 that permit conditions for naval exercises were reviewed annually and added: "We will evaluate the effectiveness of our marine mammal protective measures in light of new research findings."[129]

M. The area had just been re-opened in October 2018 when the Canadian navy and the US Coast Guard arrived to conduct live fire exercises in those same waters designated as a critical habitat. Residents cannot believe they are not allowed to go fishing in even the smallest boat when the navy can set off phosphorus bombs and use 50 caliber guns, and return 20 or 30 times a year. The navy says there is no evidence that the exercises impact the salmon or the whales, but they previously agreed not to fire munitions while whales are within roughly half a mile from ships.[130]

N. Subsea cables are increasing worldwide in number, capacity, and extent, not only due to growing demands for electrical power and telecommunications, but also owing to offshore wind farms. Modern cable sheathing retains the electric field but the magnetic field is emitted into the surrounding water.

O. All twelve stork nests located within 218 yards (200 meters) of cell phone antennae never had chicks, while only one of the eighteen nests outside a range of 328 yards (300 meters) had no offspring. The electric field intensity in the first group was 2.36V/m (±0.82), in the second group 0.53V/m (±0.82).

P. The problem also applies, to a lesser extent, to microwave ovens. Their radiation can be measured far beyond your home.

Q. The current 3G standard operates with frequencies between 800 and 2600MHz and spends 10W of electricity per tenth of a square meter (m^{-2}). The 5G system is planned to work worldwide with frequencies in three main spectrum bands: at 700MHz, at around the 3.5GHz mark, and in the 26GHz band (5g.co.uk). It is bewildering to see governments privatize frequency bands without knowing at all beforehand what effects crisscrossing permanent long-term exposure will have on living organisms. It is a huge experiment on the living world. First tests of 5G stations showed harrowing results, such as hundreds of birds dropping dead from the trees in the wake of a test run in the Netherlands.[131] Obviously, some "fine tuning" at least is drastically in order. But in the office towers of governments and industry, unfettered ecstasy ruled the day in 2018 and 2019 at the auctioning of 5G rights.

R. "Hereditary dispositions" are not set in stone. Instead, genes are being activated or deactivated by interactions with their environment and the kind of lifestyle of the person or animal (this is called *epigenetics*). Thereby our (colon) bacteria have an impact on our DNA and protein generation too.[132]

S. It was long believed that an embryo had a sterile space in the uterus. Instead, maternal bacteria reach the baby through the amniotic fluid. At birth, when the baby travels through the birth canal some microscopic ambassadors of the mother's colonic microbiome can take a ride and start a new symbiosis with the new human body. Directly after birth, skin-to-skin contact with the mother is crucial, because then the tribes of skin bacteria can pass over; this is step two. During the breastfeeding period, the colon packs tiny "lunch parcels" of bacterial travelers to be taken to the mammary ducts in the breast. And the baby receives the last biome installments as a strong foundation for a life with a healthy immune system.[133]

Caesarian babies, on the other hand, are much more exposed to the hospital bacteria. After leaving hospital, the bacterial signature of babies reveals in which hospital they were born! Therefore, medics have begun looking into ways of how Caesarian babies could be imprinted with their mother's microbiome, i.e. transferring a vaginal smear onto the baby seems the most straightforward option.[134]

T. Another study compared the microbiome biodiversity in

children in Venice and in a village in Burkina Faso, West Africa. We may generalize that people in "poor countries" have more abundant microbiomes and better immune systems than people in highly "civilized" surroundings. Hygiene can be counterproductive (of course, some bugs, like salmonella, you don't want to have around).[135]

U. Their stem cells develop in the bone marrow. Then they are imprinted in the lymphatic organs (lymph nodes, thymus, spleen, tonsils, bone marrow), i.e. they learn which substances belong to the body of the organism and which are foreign. Accordingly, they then dismantle bacteria, viruses, tumor cells, toxins, foreign particles, worms, and fungi.

V. Other heavy metals: chromium, nickel, cobalt, tin, molybdenum.

Other chemicals: toluidine (in paints and pharmaceuticals), toluene (in gasoline), benzoyl peroxide (a bleaching agent) and diurethane dimethyl acrylate.

Chapter 11

Population

The scale of the human population and the current pace of its growth contribute substantially to the loss of habitat and biological diversity. Overpopulation is *not just a cause* (like the other factors in the HIPPO formula), but a *major driver and multiplier of all ecological problems*. The human expansion continues,[1] the rates of extracting "resources", land grabbing, poisoning, waste production and CO_2 emissions are ever-increasing. And the more people join the race, the bigger the demand on the natural world and the bigger the trail of toxic waste.

To clarify the relationship between the various factors involved, the renowned Stanford professor Paul Ehrlich devised the famous I=PAT equation (pronounced *ipat*):

Impact = Population x Affluence x Technology.

Our impact on the ecosphere—the level of consumption of nature and emissions of waste—is the product of three key factors: the wealth of a society (affluence), multiplied by its size (population), multiplied again by its degree of technology.

Technology can work both ways. It can be applied to make the use of energy and raw materials more efficient, and to create new, more sustainable technologies (e.g. solar power). But it also helps to extract Earth's treasures faster and cheaper which in turn increases demand, to mask their scarcity and encourage overuse, and to accelerate their depletion.

Sadly, technological increases of efficiency have frequently proved to backfire: electricity-saving fridges and freezers have led to many households now having a second pair in the basement, and the electricity-saving LED lights have spawned 24/7 illumination of places which before simply had their lights off for long intervals. Hence, affluent societies with more advanced technology would have more chances to save energy, but all too often choose to waste it in new ways, e.g. by driving a SUV which might need three times the amount of gasoline than a standard car.

Humanity reached its first billion around 1804, its second billion in 1927, and the next two doublings merely took 48 years each.[2]

1	2	4	8 billion
1804	1927	1975	2023

Please note that the length of the intervals mirrors the *speed* or *rate* of growth. Indeed it has slowed down since the late twentieth century which led people to relax about population growth; after all, "the rates" are going down. But what has decreased is only the speed; we are still gaining over 80 million people every year. It only translates into an ever smaller *rate* of growth, but the *absolute* numbers are still increasing.

The big taboo

Reasons why the public discussion about overpopulation went silent in the 1990s:

- In the early 1970s, some scientists *predicted massive famines* because population growth was beginning to outstrip food supply.[3] These predictions luckily did not occur, largely due to the "green revolution" in intensive agriculture which increased food supply by way of new crop varieties and new (oil) technologies. The success of industrialized agriculture cast doubt on the concept of "carrying capacity" of the Earth and brought forth the belief that human numbers are not constrained by natural limits, or can at least defy them by means of new technologies.

- The new dawn of a belief in technology and human superiority coincided with the end of the Cold War in 1989, when the collapse of communism triggered a massive boost of self-confidence among the neoliberal forces in the free market economy. With the foundation of the World Trade Organization (WTO) in 1994 and the advent of globalization, the *social* market economy began to be transformed into unregulated and arrogant hypercapitalism. *Since then, the new ideology and omnipresent mantra is GROWTH.* And the growth of markets and profits also requires a growing consumer base. Neoliberal dismissal of population concerns usually accounts for the last fifty years as a huge success in (population and economic) growth but entirely ignores the devastation this growth has wreaked on the entire ecosphere. Only by denying this they can claim that the Earth, presently being overwhelmed by the needs of 7.8bn people, could even feed 15bn. Warning voices are still being dismissed as overpopulation apocalyptics or "population catastrophists".[4]
- The political left, strangely, has been on the same side as big business on the matter of population growth. Because the industrial countries have done the most to ruin the ecosphere and the climate, the left claimed that the root of environmental problems was the consumption patterns of the rich, not the growing numbers in poor countries. This accusation is fair and true, but this argument created a dualism in political

discussions which is outdated now: the concept of the rich Global North and the poor global south does not apply anymore since threshold countries like China, India and in South America have grown vast middle classes. (On the other hand, poverty is spreading fast in the fading democracies of Western hypercapitalism.) We need to tackle both everywhere: overpopulation *and* overconsumption.

- Other voices that turn the subjects of population growth into a serious taboo come from the proponents of anthropocentric humanism. Because population planning or management could easily derail into coercive population control, a colonialistic attitude towards poorer countries, and inhumane policies suppressing fundamental human rights, it is claimed to be a no-go zone.[A] However, firstly, the implication is wrong that it could not happen in accord with human rights (see main text). Secondly, it denies the fact that human population pressure is already causing *fundamentally inhumane* torture, disruption, obliteration, and annihilation of other species.[B] *A true humanism includes all living things.* And even a starkly anthropocentric take on humanism should ponder this: what could be humane about letting a multiplying humanity run over the edge, into a future without enough fresh water and food, or a benign climate. **There is no justice on a *dead* planet.**
- Green parties and environmental organizations shun the subject too, out of fear of not being politically correct. Even the World Wildlife Fund (WWF), Greenpeace, and Friends of the Earth avoid at all costs the subjects of human population and migration.

However, by not addressing it in a responsible, ethical and humane way, we leave the field to the political far right who stir fears of genocide, eugenics and population control. Rightwing Christians throw their dislike of contraception and family planning into the brew, and others (like the growth-obsessed neoliberals again) add warnings of Nazi racial theories. It all becomes very uncomfortable. And yet all fears about this subject can easily disperse in the light of facts and numbers, supplied by the UN, FAO, and a growing number of scientific studies. The good news is, *there is a humane, sustainable—and even politically correct—way of addressing the challenge of growing human population.*

All too often, population discussions focus on the high birth rates in the Global South. But large family sizes in poor countries are the direct result of high food prices and social instability which are largely caused by the globalized extractive economy controlled by the Global North. Neocolonialism has turned all natural "resources"—including human labor and well-being—into commodities to speculate with and profit from. *To change this would be the primary step towards halting human population growth and ecological overshoot.*

The other step, no less important, is to end the disempowerment and abuse of women worldwide. We need to focus on *educating girls, empowering women, and making family planning tools and information universally available.* "When societies emphasize these three pillars of civic engagement in the context of a universal human rights framework, birthrates typically fall quickly," say conservationist Tom Butler and human rights campaigner Musimbi Kanyoro.[5] A study published in April 2017 in *Science* magazine agrees that "wherever human rights-promoting policies to lower fertility rates have been implemented, birth rates have declined within a generation or two."[6] Such policies include:

- prioritizing the overall education of girls and women,
- prominent public discourse on the issue,
- establishing accessible and affordable family planning services,
- provisioning modern contraceptive methods through diverse outlets,
- making counselling for couples available,
- eliminating governmental incentives for *large* families,
- making sex education mandatory in school curricula,
- deploying health workers for grassroots education and support,
- including good healthcare for giving birth, for babies and children.[7]

Let it be said that all points around family planning are *advisory only; all decisions belong to the women and/or couples*. The problem with pressure occurs on the other side: social norms and ethics, disapproval of partners, patriarchal force, religious pro-natalist values. All of these can go as far as blackmailing and physical violence towards girls and women.

The most noteworthy point in the above list is **female education**. "Wherever women are empowered educationally, culturally, economically, politically, and legally, fertility rates fall." Studies have shown that "the number of years of a girl's or woman's education, on average, varies inversely with the number of children she will have."[8] A striking statistic from Africa shows that African women without any education have an average of **5.4** children; women who have completed primary school have one child less (**4.3**); those who finished secondary school have **2.7**; and those who advance to college

2.2. Achieving more of the other fundamental human rights for women will easily drop the average fertility rate to or below **2.1** which is the universal statistical "replacement value" (to offset people dying) to sustain a population at a given level. **The global solution is to achieve full gender equality.** Full stop.[C]

As it is, we live in a world where the women who have an average of over five children, actually give birth to eight or nine, due to infant mortality. More than just a few girls have their first child at 12, and by the time they reach 18 they are third-time pregnant. Nevertheless, the Catholic Church campaigns against artificial contraception, and Islam often favors young brides and large families. These religious views are shared by conservative political allies in the respective countries. Birth rates are largely due to social norms. *But social norms can change.*

Family planning centers are also crucial for giving help with unintended pregnancies.[D] The proportions for these, in 2012, were over 50 percent in North and South America, 45 percent in Europe, and 35 percent in Africa.[9] If we continue (monkey) business-as-usual, the UN projections expect a world population of 9.7 billion by 2050 and about 11 billion towards the end of this century. That's the optimistic lower-end calculation; it could be as many as 16 billion by 2100.[10]

To meet the expected population growth, it is estimated that food production will need to increase by about 70 percent by 2050, and double or triple by 2100. Is that even possible? The global food system is totally unsustainable in terms of pollution and soil degradation, and nearing its collapse (see Chapters 10 and 12). Additionally, global heating is bound to interrupt food production and even destroy harvests.[E] This year, we still could feed 8 billion people, but can we next year? Or in ten years?

Freshwater scarcity is another huge problem, and intensifying fast. The UN expects that water stress will potentially displace as many as 700 million people (Secretariat of the United Nations Convention to Combat Desertification, 2014). "Water scarcity, driven principally by population growth and economic growth, is set to be worsened by climate change and is thought to be a major driver of armed conflict, particularly in Africa."[F,11]

Despite the unfettered belief in technology and endless growth, the **carrying capacity** of any territory, or of planet Earth as a whole, is still a reality and a Gaian law. When the lions eat too many gazelles, their population will have to shrink too. Regions low in "resources", prone to drought and susceptible to extreme weather, like the Sahel zone or the Horn of Africa, cannot sustain as many people as richer soils in richer climes.[G] But industrialized nations live over their means on an entirely different scale: if 7 billion people lived with a total consumption level (food, clothes, water, energy, etc.) similar to that of France or the UK, it would require *two and a half Earths*; the US American way of life even four.[12] So if we say we want social justice for every human being and life standards like those of Britain, the natural limit for that would be a world population of 2.8

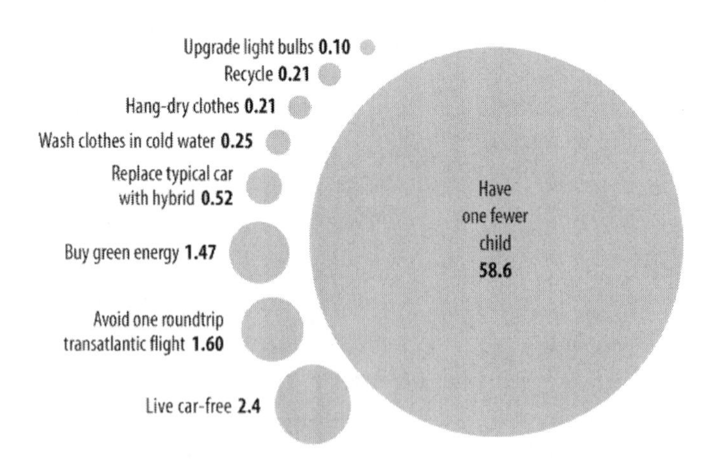

Figure 8. Personal carbon footprint: having one fewer child
In July 2017, *The Guardian* compared the CO_2 footprint of our way of life with that of our very existence.

billion—but that's still the *maximum* strain on the ecosystems so we'd better halve that: 1.4. Indeed, a number of ecologists who have looked at this from various angles have said that *a world population between one and two billion would be the truly sustainable limit for a healthy planet with healthy people.*[13]

What can I do?

- **Never ever** look with disapproval or in judgment at a mother, or a child, or any family. Talking politics is one thing, meeting real people in life is another. We can *discuss* numbers and what effects they have globally, but *meeting* another soul requires mutual respect and compassion (with the exception of dangerous psychopaths, in that case: get out of there). **Every human being on Earth has as much right to be here as you do.**
- Color differences of skin, eyes, hair, smartphone case or screen saver have no meaning whatsoever. (Unless someone has a sunburn, then give them aloe vera gel.)
- Practice empathy, for whales, for trees, for humans, for every living being.
- Find more information on populationmatters.org and overpopulation-project.com
- Watch the world population grow on www.worldometers.info/world-population
- Check out the immigrant solidarity declaration of 350.org
- "We need to have empathy now with those who are suffering [...] that's where we'll all be very shortly if we don't change course." (Mary Robinson, UN High Commissioner for Human Rights)[14]
- To reality-check your emotional self, watch this music video on YouTube: *Delta Moon – Refugee (Lyric Video).* If you don't feel moved to tears at some point, your heart must be closed and you should see a therapist.

Migration

"Refugees and economic migrants [...] are the direct result of overpopulation, ecosystem collapse, climate change, militarism and inequity. Mass migration has the potential to overrun entire societies and human civilization, and even threatens a collapse of the ecosphere. Migration must be controlled; and refugees and economic migrants assisted to return to productive, sustainable uses of land as close as possible to their place of origin," says Dr. Glen Barry who initiated the ecological sustainability website EcoInternet.[15] More people than ever before are on the move. The numbers are rising. And the percentage of climate refugees among the homeless is also rising. But so far, many mainstream media have been playing down the climate aspect of regional dramas. A prime example is Syria. There was the war and the IS, of course. But the buildup to the disaster included increasing dehydration in an already quite arid country where the population had doubled in the three decades prior to 2015. The worst drought in Syrian history[16] led to 1.5 million internal refugees before the socio-political powder keg ignited.[H]

When the large caravan of migrants from Guatemala, Honduras and El Salvador proceeded towards the Mexican-US border in October 2018, the media reported that violence and poverty in their home countries were the reason. What most reports did not say is that farmers became poor and ended up in violent slums reigned by corrupt gangs only because their crops kept failing in the first place, due to global heating.[I] If the rates of rainforest destruction in Brazil continue as they have done in the last forty years, this ecosystem is feared to reach its tipping point and collapse between 2040 and 2070 (that is without President Bolsonaro opening the rainforest for unfettered land grabbing). Already, the rainfall rates in the western parts are seriously reclining, indicating the faltering of the system. With the collapse of the Amazon, a whole continent would lose its water and could go barren within just half a decade. Africa is drying up too[J] and its population is expected to rise from 1.25 billion today to over 4 billion in 2100.[17]

The current outlook is that migration will remain a big issue for some time. In 2011, net migration contributed 68 percent of Europe's population growth. In the UK, net migration plus births has accounted for 85 percent of population growth from 2000 to 2017.[K,18] The country had reached a limit and was seriously struggling with infrastructure, rent as well as house prices, and debt and poverty levels—even well before COVID-19.[L] Unfortunately, immigration from poorer countries to richer ones plays a significant role in contributing to environmental problems: countries growing in population have more trouble meeting the goals to reduce their global footprints. In Germany, for example, a net immigration of 3 million people between 2012 and 2016 correlates with the country markedly missing its environmental targets.[19] Instead of reducing urban sprawl (land conversion of agricultural areas or forests into settlements and roads) it has grown by 155 acres (63 hectares) a day.[20] And the efforts to conserve local flora and fauna miserably failed too: the official "indicator" of species diversity and landscape quality declined to 69 percent of the 2020 target value.[21] *We have to realize that if a country wants to welcome immigrants on a large scale it has to at least double its conservation and climate efforts.*

Something else to contemplate is the present *population density* (people per square mile): Australia 9, Canada 10, USA 87, France 319, China 377, Germany 603, UK 725.[22]

A new approach is presented by the environmentalist Colin Hines: "progressive protectionism"[M] as a green case for controlling borders: "restricting not just migration but the free movement of goods, services and capital where it threatens environment, well-being [also via the 'import' of pathogens and pandemics] and social cohesion." (*The Ecologist*)[23] Note that this has to go hand in hand with a new era of regulation for banks and global trade (which is why industry and governments tend to avoid these questions and tend to keep immigration as a *purely social* matter). Strengthening "local economies" puts *people before corporations*, and saves on the large ecological footprint of long-distance shipping of goods.

Footnotes

A. "Center" politicians desperate to counter rightwing populists who strongly oppose immigration often say that it is good for the country to add to the work force. They pretend to play the humanity card but actually voice the desires of the industries for cheaper labor (even with white collar jobs they'll often save money). Modern immigration policy pretends to be humanistic, but is dominated by the economy: people as a resource. Too often, development aid is used as a political instrument instead of creating incentives to ensure that valuable skilled workers remain in the countries most in need of development (or re-development after being bombed to shreds by Western nations). Is it humane to steal skilled workers from poor countries which originally paid for their education?

B. An example from history: claiming the Great Plains for food production could only happen because 99 percent of the native biota had been wiped out. The grassland ecosystem with its great diversity of plants, animals and other organisms had been sacrificed, and, in the 1870s, over 100 million bison were slaughtered.[24]

C. Liberating women and increasing gender equality and thereby creating higher education levels have positive repercussions on national economy and affluence, as can be seen in countries which have for some time applied some of the policies mentioned, for example, Thailand, Costa Rica, and South Korea.

Reversing population growth is "an opportunity for

government policies and humanitarian health programs operated by NGOs to focus on actions that are effective and morally just, and which have profoundly positive effects for individuals, families, nations, and the Earth."[25]

D. "Each year, about 85 million unintended pregnancies result in 32 million unplanned births worldwide"[26] that resembles about 40 percent of the annual growth rate.

   Avoiding unintended births would have disproportionately positive effects in wealthy countries too: "For example, stopping population growth in the U.S., where roughly half of all pregnancies are unintended, will have far greater climate and environmental benefits than, say, lowering Niger's total fertility rate, which is more than seven children per woman. The former will help reduce negative pressures on the global biotic community, the latter would help lessen suffering of families and local communities in one of the world's poorest countries."[27]

E. The Intergovernmental Panel on Climate Change (IPCC) warns that global heating may have severe impacts on food security via higher temperatures, precipitation changes, increased frequency of extreme weather events, the spread of new pests and ocean acidification.

F. Another factor is work. A global trend towards higher unemployment and underemployment has been emerging for years. This has particularly hit younger people, and the costs and risks of population growth are also generally worse for younger generations than for the older ones. The oncoming digital revolution which will make millions of jobs in each country redundant is bound to further significantly diminish people's ability to earn a livelihood and live their lives in dignity.

G. And it is in the fragile zones that nature is easily overexploited; for example, subsistence farmers or goat herders using the sparse tree or shrub cover for fire wood in regions bordering the Sahara have driven the continuous spreading of the desert over the last decades.

H. Demonstrations against the Syrian government, or rather against poverty, began in 2011. Four years later, almost half a million people were dead and 5 million became emigrant refugees. The country that had long passed its carrying capacity expelled about a quarter of its population. President Assad said in 2017: "We have gained a more homogenous, healthier society."[28] That is one way of dealing with overpopulation!

I. Robert Albro, a researcher at the Center for Latin American and Latino Studies at American University, said in 2018: "The main reason people are moving is because they don't have anything to eat. This has a strong link to climate change—we are seeing tremendous climate instability that is radically changing food security in the region."[29]

J. One example: On the southern edge of the Sahara, Lake Chad used to provide a lifeline to millions of people in Nigeria, Cameroon, Chad and Niger. But it has shrunk by about 95 percent. With the disappearance of its water, the carrying capacity of the entire region has decreased dramatically.[30] This, however, did not stop the government of Chad in 2020 to ask to suspend the application for World Heritage Site status for Lake Chad, planning to unleash oil and mining companies to explore the region.[31]

K. At that current rate, the UK population was expected to rise by nearly 8 million people in the next fifteen years.[32] Hence the British focused on immigration in the Brexit talks.

L. Over 500,000 people in 154 countries were asked between 2010 and 2012 about their future dreams. 13 percent of the world population (630 million people) "say they would like to leave their country and move somewhere else permanently." The top desired country was the USA with 138 million aspirants, followed by the UK with 42 million.[33]

M. "Progressive protectionism aims to nurture and rebuild local economies in a way that permanently reduces the amount of international trade in goods, money and services and enables nation states to control the level of migration that their citizens desire." (Hines 2017)

Chapter 12

Overconsumption

> We have in fact gained a world of resources by forfeiting the living Earth.
> – Eileen Crist[1]

Humans eat other living beings. This is not unnatural. Humans have animality, and all animals eat other animals or plants. In ecological terms, there are *autotrophs*, such as plants and algae, that manufacture their own food via photosynthesis, and there are *heterotrophs*, such as humans and other animals, who eat autotrophs or other heterotrophs, or both. A third group are the *detrivores*, such as certain fungi and bacteria, that break down dead organic material and thereby release again the nutrients for the first group.

The problem today is the sheer numbers of humans and their unbalanced diets which often demand more than necessary. Mass feeding requires mass production. Killing for food has become big business. 300 mammal species are being hunted and eaten into extinction, while the oceans are massively overfished, and agribusiness demands ever more land from the wild. A 2016 study showed that of the near-threatened or threatened species on the IUCN's Red List, 72 percent of the species were affected by overexploitation: logging, hunting, fishing or gathering them from the wild. To biodiversity, agriculture and overconsumption still are a bigger imminent threat than the climate emergency[2] (although the latter is catching up fast).

Runaway human consumption of "resources" devours wild nature habitats and exhausts the abundance of species. To communicate this, Earth Overshoot Day (overshootday.org) was invented as the (fictive) date when humanity's demand for ecological "resources" and services in a given year exceeds what Earth can regenerate in that year. In 2019 it reached 29 July, while in 2020 corona broke the trend and took it backward to 22 August. However, the amount of material consumed annually by the global economy has quadrupled since 1970, and in 2019 reached an all-time high by passing the 100-billion-tons mark. That's hundred times a thousand million metric tons. Of these, just over half was minerals, a quarter was crops and trees, 15 percent fossil fuels, and 10 percent ores. Altogether, only 8.6 percent of materials was recycled in the same period.[3]

Mining for minerals and metal ores is a major driver for habitat and species loss. Mining for gold is particularly disastrous. It releases highly toxic mercury which poisons entire areas and rivers long-term. Additionally, many heavy elements such as cadmium, lead, zinc, copper, and arsenic which occur close to or intermingled with gold ore, pass into surface water and groundwater, a process called acid mine drainage. Such toxic dumps are long-term, highly hazardous, and considered second only to nuclear waste dumps. Gold extraction is also highly energy-intensive (as is the production of cars, devices, gadgets, and all sorts of trivial "stuff"). Gold mining, especially illegal mining, has increased throughout South America since the financial crisis began to push gold prices higher in 2008. In the Madre de Dios region of the Peruvian Amazon alone, there are 50,000 to 70,000 illegal gold prospectors who use over 90 metric tons of mercury every year. Since much of it ends up in rivers and most locals feed on fish, 90 percent of gold diggers have up to 3 times the WHO limit of mercury in their blood, Indigenous people up to 35 times. Peru alone exports 170 metric tons of gold annually, mainly to Europe. In the Amazon region of Madre de Dios alone, 172 acres (70 hectares) of rainforest have disappeared in just a few years.[4]

Another example of the way runaway human consumption adversely affects water and the living world is the damage done by the countless jungle labs in the drug trade. Coca leaves are powdered with cement, soaked in gasoline, and then extracted into a coca base with the help of battery acid. Thus jet fuel, ammonia, hydrochloric acid, sulfuric acid, acetone and calcium oxides are among the substances that are frequently just tipped down a hillside.[5]

Water

The Earth is covered with water. But 97 percent is salty, 2 percent is locked up in ice and glaciers, and some water exists as moisture in the atmosphere. Less than one percent is available for all creatures in the ecosphere. Water is so essential to living organisms that the Gaia pioneer VI Vernadsky aptly called life *animated water*.[6] Water is precious.

For millennia, human societies have lived with a modest footprint, and their water use would be restored by natural cycles of respiration, evaporation, and rain. But since the Industrial Age and the population explosion, with their ever-increasing water demand for factories and agriculture, the ability of freshwater to restore and sustain itself is being outrun in most parts of the world. Often, the countries with the highest fertility rates are also the driest ones. In this situation, the contamination of freshwater (e.g. with pesticides from fields and golf courses, nitrate fertilizers, radioactivity, chemicals, plastics, gasoline and oil leakages) is the peak of insanity. Or maybe this is: global corporations buying freshwater rights in remote communities for the production of soft drinks (exactly those that cause obesity and increase the diabetes risk around the world, due to their immense sugar contents). To produce one liter of Coca-Cola it takes four liters of clean water.[7]

Today, "2.1 billion people have no regular access to clean drinking water."[8] The UN expects that *almost half the world's population* will be living in areas of high water stress as early as 2030, and 5 billion people by 2050.[A,9]

The water security in a "wet country" such as England is not a given either. Since the privatization of waterworks in the UK in 1989, water bills have increased 40 percent above inflation, shareholder dividends have soared upwards of £13.5bn in annual payouts, and the profit-driven companies refuse to invest in plugging leaks which lose 3 billion liters of water a day. Count in global heating and population growth, and England is expected to run short of water in the mid-2040s. The necessary step would be to nationalize water (and other fundamental services such as schools, transport, health systems, mail) again, and repair the leaks.[10]

What can I do?

- Avoid products that take a high toll on rainforests or water. Check out your water consumption footprint:[11] https://www.watercalculator.org

  Further info on water footprint: https://www.treehugger.com/clean-technology/how-many-gallons-of-water-does-it-take-to-make.html
- Watch these little info videos about overconsumption, preferably with your children: storyofstuff.org/movies, about water. "Our Water, Our Future": storyofstuff.org/movies/our-water-our-future
- Visit We Own It, an organization that campaigns for "public services for people not profit": weownit.org.uk

Fishing

In many developing countries (and "transition" countries), fish is the major source of protein and essential micronutrients that rural people can access or afford. However, the appetite for fish in developed countries has been increasing, and for some time now their demand cannot be met by fish stocks in their own waters. Hence, large fleets of industrial fishing boats scouring the waters of developing countries have created a wholly different situation: industrially caught fish has become a globalized commodity that is mostly traded *between continents* rather than consumed in the countries where it was caught, thereby adding vastly to the impact of global transport routes requiring energy (usually crude ship oil, see Chapter 14) and contributing to climate disruption. On the other hand, the small-scale fisheries that traditionally supplied seafood to coastal rural communities (and possibly further inland as well) are forced to compete with the export-oriented industrial fleets, and without much support from their governments—not only in African countries but also, for example, in Cornwall, UK.

In the world's oceans, 93 percent of fisheries are "fully fished" or already overfished.[B] The body to monitor and supposedly regulate this is FAO, the Food and Agriculture Organization of the United Nations. But proof emerged in 2016 that FAO—by explicitly excluding the amounts of discarded bycatch, and illegal and otherwise unreported catch—is *considerably underestimating the grand totals*. After all, FAO can only report what its member states are telling it. A 2016 study published in

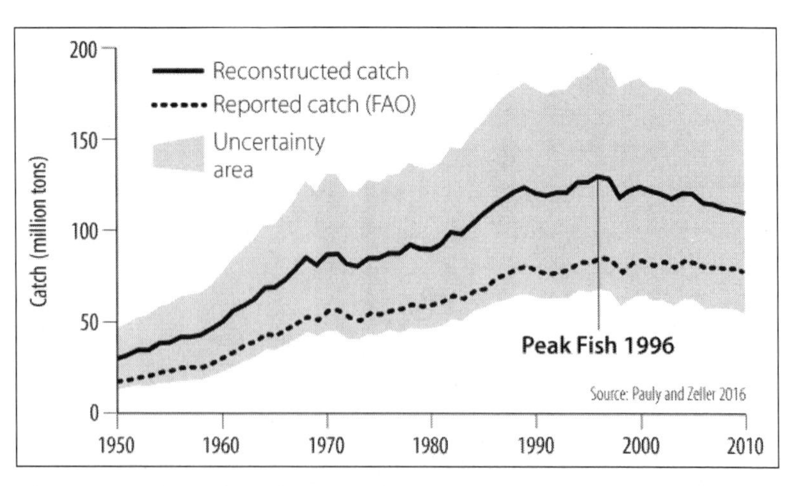

Figure 9. Trajectories of reported and reconstructed marine fisheries catches 1950–2010

Nature reports that the real catches are overall 53 percent higher than the reported data.[C] It is clear now that after Peak Fish in 1996 more and more marine zones are becoming overfished.

Tending to make things worse quicker, the WTO is still beating around the bush to ban fishing subsidies. Subsidies paid by governments to the fishing industry amount to around US$35 billion per year, of which 20 billion are given as fuel subsidies and tax exemption programs that particularly put large fishing fleets at an advantage.[12] And it is exactly these which also cause another major problem, namely "collateral damage" and bycatch.

Two examples for collateral damage and bycatch. The numbers of the North Atlantic right whale (*Eubalaena glacialis*) had been recovering since a moratorium on its hunting in 1935, and by 2000 it was thought that there were an estimated 400 right whales in the North Atlantic. But in 2017 it became clear that their numbers were plunging again, with perhaps only around 100 reproductively mature females left in the sea. The reason being the use of increasingly heavy commercial fishing gear which is dropped onto the seabed to catch lobsters, crabs and other creatures off the North American east coast. Whales swim into the rope lines and become entangled. In some cases hundreds of meters of heavy rope, complete with traps weighing more than 60kg, have been found wrapped around whales who often die of exhaustion carrying these weights around.[13] In July 2020, the North Atlantic right whale was moved from Endangered to Critically Endangered on the IUCN Red List, that means: one step from extinction.[14]

Climate disruption also plays a part. The North Atlantic right whale's primary feeding ground, the Gulf of Maine, has warmed three times faster than the rest of the world's oceans in the period 1990-2020. The weakening Gulf Stream brings less cold currents into the Gulf of Maine and the tiny crustaceans called copepods, the staple food of the whales (along with krill, sea snails and sea slugs), move north to cooler regions.

Populations of the iconic leatherback turtle (*Dermochelys coriacea*) have declined by 97 percent over the past three decades. They are now critically endangered, mostly due to long-line fishing in the high seas. Some trawlers drag fishing lines that are more than 75 miles long (!), and tens of thousands of sea turtles get snagged on the hooks and drown every year.[15] Turtles, whales, dolphins, fish, seabirds, deep-sea corals, and whole ecosystems suffer because there is no protection at all for species, endangered or otherwise, on seas outside national waters. Over 70 percent of our planet is covered with ocean, and only 7 percent of its area protected.[16]

And as for fish farms, the pools are rife with chemicals and antibiotics. And we should never forget the ethics either: fish are conscious, sentient beings, they have both memory and a capacity for suffering. Fish have feelings too and we need to include them in our moral circle.[D]

What can I do?

- Choose only fish products which are dolphin-friendly.
- Choose only tuna products which are tuna-friendly, in other words: none (or as little as possible).
- Leave sharks alone too.
- If you want to cut down on your own fish consumption you must make sure to give your body enough omega fatty acids and vitamin D, particularly when you live in a temperate country with lack of sunlight. The best vegan source of omega-3 and omega-6 is hemp oil.

Agribusiness

At least the chemical pollution caused by industrial agriculture is present in the public debate. But the initial problem is the vast amount of land required, resulting in land grabbing and therefore habitat destruction worldwide. And the third concern, almost entirely ignored, is the serious matter of **soil degradation**. As *The Guardian* columnist George Monbiot says: "To judge by its absence from the media, most journalists consider it unworthy of consideration. But all human life depends on it. We knew this long ago, but somehow it has been forgotten. As a Sanskrit text written in about 1500 BCE noted: 'Upon this handful of soil our survival depends. Husband it and it will grow our food, our fuel and our shelter and surround us with beauty. Abuse it and the soil will collapse and die, taking humanity with it.'"[17]

The other root of soil degradation—apart from poisoning the soil biota with pesticides, pharmaceuticals, and synthetic fertilizers—is mechanical: ground compaction by (super-)heavy machinery doesn't only crush cute and essential earthworms and bugs but squeezes out the oxygen and alters ground chemistry.[18] And worse: (at least in northwestern Europe) failing to seed winter ground cover lets the winter winds and precipitation carry away the top soil, leading to erosion of the fields, clogged up riverbeds and, with increasing irregularity, floods downriver. This is particularly bad in areas with maize crops; and because maize is in such demand for biomass power plants, many regions in Europe don't grow food anymore but up to 70 percent maize. A single maize field can lose ten or twenty metric tons of soil per year. These areas are prone to desertification in the near future.[19]

With the loss of soil we don't mourn only the fading abundance of benevolent creepy-crawlies, worms and microorganisms. *The soils of the world are also a huge carbon reserve, the top meter alone contains three times as much carbon as the entire atmosphere.* But due to soil degradation caused by farming practices, 133Gt (billion metric tons) of carbon have been lost from the top two meters of the world's soil since the dawn of agriculture. This figure is known as the total "soil carbon debt", and represents a mean loss of 8 percent of the original carbon reserve, according to research published in the *Proceedings of the National Academy of Sciences*.[20]

Industrial farming over the last hundred years has increased the rate of soil erosion *sixtyfold*. Already in January 2014, the UN announced that *a third of the world's soils have now become "severely degraded"*[21] and therefore unusable for agriculture. We are now working towards half. "Landowners around the world are now engaged in an orgy of soil destruction so intense that, according to a senior FAO official, the world on average could have just 60 more years of growing crops."[22] But the UN can only advise. Some local governments, as in the UK in 2015,[23] revise their soil standards, but so far without penalties or serious consequences.

The good news: There are already methods to get the lost carbon back into the soils. An increase of just 1 percent of the carbon stocks in the top meter of soils would bind more atmospheric carbon than the annual CO_2 emissions from fossil fuel burning, says the International Union for Conservation of Nature (IUCN), and continues: "Many innovations in sustainable land management are now known and recognized for their multiple environmental, social and economic benefits."[24]

Regarding pasture land, *managed grazing*, which involves moving cows around to graze on alternating patches, is an effective way to sequester carbon,[25] because it avoids overgrazing, thus giving the root mass a chance to recover quicker and bind more carbon. For plantation land there are methods like *pyrolysis* which transforms biological carbon into a mineralized form that doesn't rot and that stays in the ground. This charcoal counteracts acidification, replenishes nutrient contents and improves the soil flora and fauna. Charcoal also filters water, which is of inestimable value. In these organic ways, it would be possible over the next few decades to bind 140 billion metric tons of carbon in the ground again[26]—that would almost solve the climate crisis.

What can I do?

- Go organic: food, cotton, everything.
- Spread the word that the first reason for organic farming never was healthier food but healthier soil and land.
- Participate in farmers' markets.
- Check out Community Supported Agriculture (CSA)[27] which is an ingenious way of building new regional infrastructures for people instead of machines and corporates: communitysupportedagriculture.org.uk

Livestock

Rearing livestock causes more greenhouse gas emissions than all the world's vehicles, trains, ships and planes combined (see Chapter 14). The world has wasted four decades discussing fuel efficiency in the transport sector (with not much showing for it either), but the biggest culprit for the destruction of the living world (including the climate) is industrial agribusiness in general, and the meat sector in particular.

To start with, meat production has a tremendous demand on water. A 2010 study of water footprints states that while vegetables on average need about 322 liters of water per kg (2.2 pounds), and fruits 962, meat is a different ballgame: chicken 4,325 liters per kg, pork 5,988, sheep/goat meat at 8,763, and beef 15,415 liters per kg. Agribusiness accounts for about 70 percent of water used in the world today, but a 2013 study found that it uses up to 92 percent of our freshwater, with nearly one-third of that only for livestock rearing.[28]

Let us not take this out on the few domesticated animals of small farmers in poorer countries where meat, eggs and dairy are important in providing nutrients, especially for children, and also provide leather, wool, manure, transport and plough pulling. This is about Big Meat. Costco, for example, a company which is setting up a new plant in Nebraska to process more than

2 million chickens a week, or more than 100 million a year. It is precisely the sheer size of factory farming and mass production of cheap meat that account for their massive impact on climate disruption, and puts a hoofprint on the land so gigantic it can be detected from outer space: in the USA, just 4 percent of the land surface (77.3 million acres) is used to grow food for humans, but 34.6 percent is cattle pasture and 6.7 percent is used for growing livestock feed. All in all, *more than ten times* (41.3 percent) the amount of land is used to feed livestock than to feed humans directly.[29]

The industrialized production of meat (and dairy and eggs) has led to a situation where the balance of biomass is completely out of hand. With the ongoing sixth mass extinction and the loss of biosphere abundance, humans who once were *just one* mammalian species among hundreds, now make up 36 percent of all land mammals. And—lo and behold!—domesticated livestock (mostly cattle and pigs) comprises a further 60 percent. That's 96 percent humans and their red meat, leaving just 4 percent wild mammals. Similarly, of all the birds in the world 70 percent are now chicken and poultry in their death camps.[30] Overall, already more than 90 percent of farmed animals

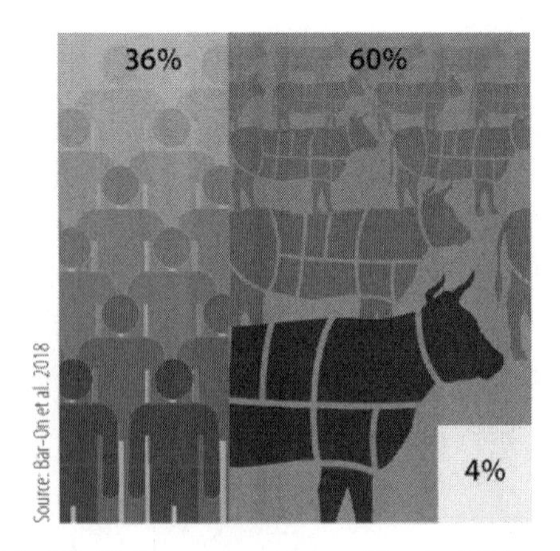

Figure 10. Of all the mammals on Earth, 96% are livestock and humans

globally are living in factory farms.[31]

Even the maths don't add up: *global meat (and dairy*[E]*) production requires 83 percent of all farmland, but only produces 18 percent of our food calories.*[32] Just to account for the climate damage, meat prices would need to be three times as high (based on the—questionable—international system of carbon offset certificates). The fact that they aren't is due to a) generous subsidies (despite its large-scale destruction of wild nature and harm to the climate and to workers, agribusiness worldwide receives one million dollar in subsidies—per minute[33]), and b) the legal setup that "companies can just externalize the costs of their [also epidemiologically, see 'Pandemics' in Chapter 9] dangerous operations on everyone else," as biologist Rob Wallace sums it up. "From the animals themselves to consumers, farmworkers, local environments, and governments across jurisdictions. The damages are so extensive that if we were to return those costs onto company balance sheets, agribusiness as we know it would be ended forever. No company could support the costs of the damage it imposes."[34]

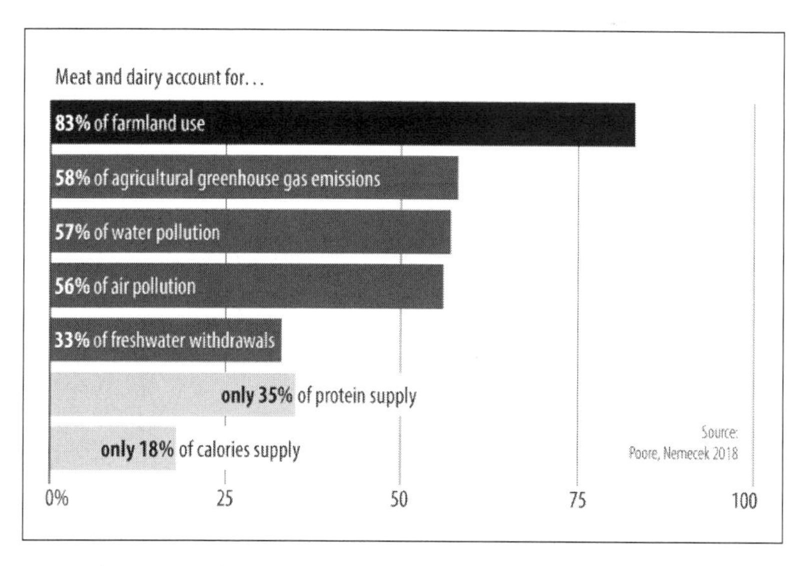

Figure 11. The hidden toll of meat (and dairy) production

Red meat isn't even healthy. At all. In 2015, the World Health Organization declared unprocessed red meat to be a probable carcinogen and processed red meat to be a carcinogen, placing it in the same category as asbestos, arsenic, alcohol, and tobacco.[F,35] Apart from colon cancer, there is also a link to mental health problems.[36] Still, many people in rich nations eat more than the recommended amount of red meat, which is linked to heart disease, strokes and diabetes. Globally, an estimated $285bn are spent every year treating illness caused by eating red meat. If this spending in national health was to be fairly incorporated into the meat prices, bacon in the UK would need to be taxed at 80 percent, in Australia at 110 percent, and in the USA at over 160 percent.[37]

In the Amazon and neighboring ecosystems, deforestation and land grabbing continues at terrifying speed. The mostly illegally felled timber is sold worldwide (Europe is still one of the biggest importers[G]). But the main reason for clearcutting is cattle pastures and fodder crops. In Brazil, over 1 million

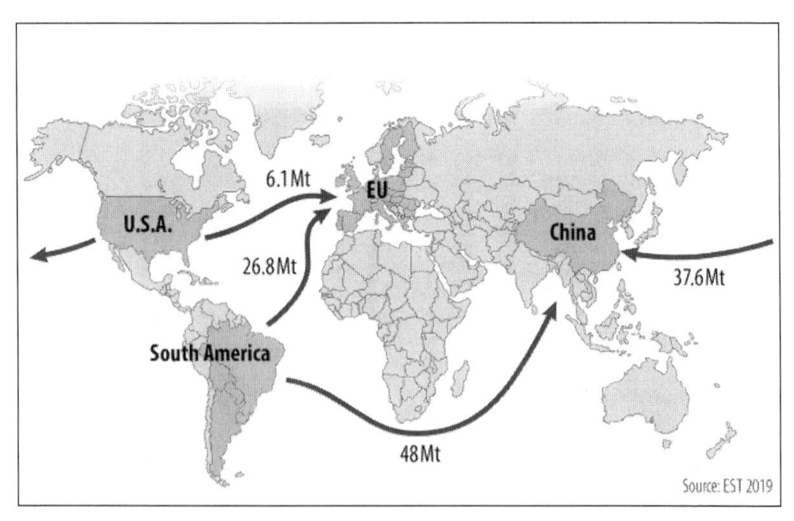

Figure 12. The global soy trade: tropical rainforest for meat

European countries are the biggest importers of South American soy products, after China.

square miles have been deforested for agribusiness,[38] an area more than ten times the size of Oregon or the UK. Genetically modified (GMO) soy and glyphosate are ruling the day,[H] polluting groundwater, rivers, lakes, children of man and beast. Indigenous villages are being pushed out, often violently. Also in Bolivia, Argentina, Paraguay. And in Indonesia for palm oil. The global agro business—soy, palm oil, cacao, grains—is mostly in the hands of only five corporations,[I] with an annual multi-billion-dollar turnover. South American soy fodder has two main destinations: China and Europe. The USA meets its own demand and has a surplus that also goes to China (or *did*, until Trump's trade war forced the Chinese to turn to Brazil, which increased pressure on the rainforest).

Insane Trade

The logic of globalization dictates that goods have to travel ever longer distances from production to the consumer. Extra miles create extra GDP. In the USA, any food product has traveled an average 1,500 miles from its place of origin to the consumer.[39] Tuna fish from the US East Coast is flown to Japan to be processed, then flown back to the States. In the UK, butter from New Zealand and apples and pears from South America, for example, often are cheaper than home produce. English apples are *flown* to South Africa to be waxed, then *flown* back to the UK. Many countries import and export similar or near-identical qualities of foodstuffs. In 2007, Britain exported 20 metric tons of bottled water to Australia, and Australia exported 20 metric tons of bottled water to Britain. All of this is possible because of lax trade regulations and subsidies systems favoring neoliberal free markets with transport and energy based on Cheap Oil. "The system we developed could not be more wasteful," says UK MEP Zac Goldsmith.[40]

USA	**imports**	**exports** (tons per year)[41]
sugar	70,820	83,083
potatoes	365,350	324,544
beef	953,142	899,834

What is being done?

The system adamantly holds on to the status quo. Olivier de Schutter,[J] co-chair of the International Panel of Experts on Sustainable Food Systems (IPES-Food), reminds us that if food prices would include the environmental (habitat loss and poisoning) and social damage (unemployment, poverty, public health), they would need to be *twice as high*. But no short-term politician who wants to get re-elected in a few years' time can afford the resulting public outcry, so the system keeps subsidizing the biggest land owners. The UN, governments, and scientists agree that the current food system has reached a dead end (for all the degradation of climate and ecosystems it doesn't even feed all humans properly: almost 1 billion people are still hungry). But the governments' advisors are the big corporations. And the biggest ones are mineral oil companies. Cheap Oil keeps the tractors rolling; fertilizers and pesticides are made from oil derivates, and so is almost all the fuel that ships food around the planet.[K]

- *GHG release through industrial farming:* Not much. The climate discussion avoids the subject and instead focuses on (private) transport (see "The carbon discussion" in Chapter 14).
- *About antibiotics:* A small reduction in the EU by 2022.[L]

- *Towards plant-sourced protein:* In terms of ecology as well as public health, mass meat production is the single biggest folly of humanity. This is why scientists have suggested that the UN climate conferences (COP) should urgently take up "animal to plant-sourced protein shifts" as "part of countries' mitigation commitments."[42] But no news about this got to the press after COP24 was held in December 2018.
- Many people don't want to just wait for a better world to come. A 2017 study showed that if all Americans substituted beans for beef, the country would be close to meeting the Paris GHG goals agreed by Barack Obama.[43] Or, if Europeans and North Americans would cut their red meat consumptions by 80 percent, 60 percent of today's pasture could be given back to wild nature.[44] And the insight that "giving up beef will reduce carbon footprint more than cars"[45] (see "The carbon discussion" in Chapter 14) could have been spreading at least since 2014, but media and the authorities rarely ever touch the subject. Instead, we have seen the advent of big advertising campaigns promoting meat like never before. Despite such efforts, numbers of vegetarians and vegans are rising. In Great Britain, for example, from 2006 to 2019 the number of vegans has quadrupled to 600,000, and demand for meat-free food in the UK has increased ten-fold.[46]

Food expert Olivier de Schutter conducted a UN study that concluded that organic farming can very well feed the world. Since large farms mostly produce for livestock feed and biofuels, and 70-75 percent of the world's food comes from small farms already, all it needs is tweaking their potential to improve some of their growing techniques: going organic, using agroforestry (which can increase crop diversity as well as yields), agroecology (which can double the yields of a given area size), or permaculture (which can treble or quadruple yields). Many examples around the world have long shown that small farms are the way into the future.

Prof. Dave Goulson of University of Sussex has made the case for allotments,[47] little smallhold farms or rather large gardens in which people grow their own organic food. Applying various sustainable techniques such as permaculture, agroforestry and biodynamic farming will equally benefit the health of the people and the health of the ecosystem they live in:

- Allotments have a high insect biodiversity and hence also attract birds.
- Allotments produce a lot of food, and with zero food miles, zero packaging, and uses minimal or no pesticides. Small patches of crops are much less susceptible to pests, and their natural predators (e.g. ladybirds [ladybugs], lacewings, hoverflies) are generally much more abundant, hence even if pests do find a crop they tend not to flourish.
- Allotment soils tend to be healthier than farmland soils, with more worms and higher carbon content, which helps to tackle the climate crisis. Because an allotment is never stripped bare (as happens when an arable crop is harvested), the soil does not erode and organic matter can build up over time. Perennial crops such as fruit bushes and fruit trees also protect the soil.[48]
- Allotmenteers tend to be healthier than neighbors without allotments, particularly in old age.
- Experienced allotmenteers can produce the equivalent of about 35 tons of food per hectare (2.5 acres). For comparison: industrial crops of wheat and oilseed rape produce about 8 tons and 3.5 tons per hectare, respectively.

Because of such effective productive rates, even a densely populated country like the UK could produce all its fruit and veg on just half a million acres (200,000 hectares) of land: the equivalent of 40 percent of the current area of gardens in the UK, or just 2 percent of the current area of farmland. Which means that a huge percentage of currently degenerating British farmland could be freed up for rewilding.

Funding for such a vast systemic shift in agriculture would be easy to generate. The UK currently consumes about 6.9 million metric tons of fruit and vegetables per year, and 77 percent of that is imported at a cost of £9.2 billion. And there are £3.5 billion currently given out in farm subsidies which still favor an unethical and ecocidal abuse of land, plants and animals. *There is enough money in the country to create funds to help people buy or lease allotments, receive training and free seeds.*

The most widespread doubt about this model of food production is that it is much more labor-intensive than industrial farming with its heavy machines and (increasingly expensive) technology. However, industrial farming only produces about a quarter of human food, it is predominantly dedicated to energy crops and plants for sugar production. And on the other hand, unemployment rates are high and are set to rise in most countries. The long-going demise of rural communities resulting from the industrial model of agriculture will soon even get worse because Artificial Intelligence (AI) in "smart farming" will make almost all humans redundant. And as the Digital Revolution is expected to destroy far more jobs over the next decade than the corona lockdown did temporarily in 2020, creating meaningful work opportunities in the countryside (and with little or no commuting!) is one way forward.

PS: In Russia, almost 40 percent of the food produced comes from small household gardens, the popular dachas.[49]

What can I do?

- Reducing meat and dairy consumption is the single biggest way everyone can lessen the human impact on the planet. *Balancing climate disruption is impossible without massive reductions in global meat consumption.* It won't kill anyone to strongly reduce meat and fish from the national average of 60kg/year (1.15kg/week).
- Spread the word about the unsustainability of meat production. But never try to force anyone to change diet or the direction of their thinking; that would only harden positions and backfire. Instead, challenge politicians about the global facts (which they know but adamantly ignore).
- Put pressure on your local supermarket chain to make their supply roots transparent: do their meat suppliers use rainforest soy as fodder?
- Avoid cured products such as bacon, ham and sausages, as the food additives that turn the meat pink are carcinogenic.
- Replace dairy milk with vegan options. Any plant-based milk is better than that from animals. However, plant milks (and foods) also have drawbacks. Almond monocultures (at least in California) involve large-scale bee abuse. Rice plantations produce more GHG emissions than any other plant milk, expel a lot of methane, and pollute waterways with large amounts of fertilizer. Soy is a vast global business and most of it is grown as GMO crops. Coconut plantations exploit workers and destroy rainforests in the Philippines, Indonesia, and India. And all of the above require (fossil fuel-consuming) transport along global trade routes. Hence the winners are: locally produced hemp milk and oat milk.
- Watch the documentary *Food, Inc.* (trailer on YouTube:

"Food, inc. (2008) Official Trailer #1"). The movie's official website also has offers to help you organize a public screening: http://www.takepart.com/foodinc/faq/index.html

Footnotes

A. An additional result of aquifer depletion is that using deep groundwater adds more CO_2 to the atmosphere than volcanoes do: groundwater contains from 10 to 100 times as much CO_2 as surface water in lakes and rivers (Harwood 2010). Now we finally understand what the "water wizard " Victor Schauberger (1885-1958) meant when he warned against "immature" water from pumps, and instead promoted only spring water as "ready" and "healthy". Healthy not just for humans and trees, but for the whole biosphere. But we are far from that; some pump wells in the American corn belt are hundreds of yards deep by now, and their water levels are still continuously falling.[50]

B. 60 percent of waters are being "fully fished" to their limits, and the percentage of stocks fished at biologically unsustainable levels increased to 33 percent in 2015; for the principal market tuna species it is 43 percent, according to the FAO report 2018.[51]

C. The FAO underestimation is widely known among scientists, and also is freely acknowledged by FAO itself, but its sheer global magnitude was not known before. While the FAO statistics suggest that the world catch increased fairly steadily to 86Mt (megatons = million metric tons) in 1996, leveled off, and then slowly declined to around 77Mt by 2010, the study found that "the reconstructed catch peaked at 130Mt in 1996 and declined more strongly since."[52]

D. Recommended reading: Marc Bekoff 2014. Fish Are Sentient and Emotional Beings and Clearly Feel Pain. psychologytoday.com, June 19.[53]

Also: Carl Safina 2018. Are we wrong to assume fish can't feel pain? theguardian.com, 30 October.[54]

E. Cow milk only requires 3 percent of the current livestock land use.

F. Due to the added nitrites and nitrosamines, processed meats have been linked to 34,000 cases of colorectal cancer worldwide each year; 6,600 bowel cancer cases alone in the UK—four times more than fatal traffic accidents in the UK.

G. The illegal timber imported into Europe is "white-washed" with false import/export papers via China and/or Russia. And some of the charcoal from Amazon fire clearances can also be bought in Europe, as barbecue coal.

H. In 2017, of the 85.7 million acres (34.7 million hectares) soybean plantations in Brazil, 97 percent comprised of GMO crops.[55]

I. Archer Daniels Midland (ADM), Bunge, Cargill, Louis Dreyfus, and Wilmar.[56]

J. Olivier de Schutter is a Belgian legal scholar who served the UN as an independent expert on food. This paragraph is translated from his interview in the French documentary *Tomorrow: Take concrete steps to a sustainable future,* a film by Cyril Dion and Mélanie Laurent (2015). https://www.tomorrow-documentary.com/

K. Biofuels represent less than two percent of the world's consumption of transport fuels (FAO[57]).

L. The new EU regulation bans the use of human reserve antibiotics in veterinary medicine and the use of unprescribed animal antimicrobials. The Green MEP Molly Scott Cato is optimistic about its effect on factory farming: "The restrictions on antibiotic use will also challenge the factory farming model where animals suffer appalling conditions and are packed together in unhealthy conditions [...] Without the routine use of antibiotics, farmers will need to adopt better farming practices that will improve the life of farm animals across the EU. This is a major victory for public health and for animal welfare."[58] It is not clear whether all these new rules will be fully applied in the UK after Brexit.

Chapter 13

Energy and "Progress"

Overconsumption is not just about food. By definition, capitalism is the means to maximize personal profit, and to be most effective it needs to grab everything that can be monetized (including nature and her treasures), to increase production of products with a *limited* lifespan (so they need replacement soon), and to constantly enlarge the consumer base (population). And all of this can be enhanced when cheap energy is available.

The earliest human species (*Homo erectus*) began the controlled use of fire for heating and cooking food about a million years ago. In evolutionary terms, human groups with fire had distinct advantages, and eventually they spread all around the globe. And for most of human history, their main fuel was wood (with some leaves and grasses).

At least since Roman times (a millennium earlier in China), the use of coal intensified the heat output for early industries such as glass, pottery and metalworks. These were the beginnings of the use of *fossil* fuels. But only with the growing use of crude oil and combustion engines in the second half of the 19th century an entirely new era of productivity began. As the writer Rob Mielcarski puts it so aptly, "unlike sunlight that is constrained to the real-time flow from the sun, fossil energy accumulated over millions of years and therefore acts as a giant solar energy battery. Now humans could not only exploit current solar energy (e.g. grass) and recent solar energy (e.g. wood) but also ancient solar energy (e.g. coal, oil, natural gas)."[1]

"Because energy," he continues, "is the master resource that can be used to extract other resources, including more energy, fossil energy created a positive-feedback-driven 200-year period of explosive population, wealth, and technology growth." What most people don't know: the 1960s "Green Revolution" of the global food system that has carried human population beyond the 3 billion count *has been—and is—entirely dependent on Cheap Oil*. Not only the diesel-powered tractors, harvesters, and lorries, not just the trucks and ships that deliver the produce all around the world depend on the oil trade, the whole range of chemical warfare ("pesticides") is oil-based too. But what really made the human population explosion possible is the fertilizers made of synthetic nitrates (it takes **three tons of oil to make one ton of fertilizer**). Prof. William E. Rees of the University of British Columbia points out that while 150 years ago, 100 percent of our food had essentially derived from solar energy, now 90 percent of it is a product of fossil fuel.[2] And the ecologist Michael Novack states that "it was chemical fertilizers that both made higher yields possible *and* began the attitude that we were not limited by Nature."[3]

The fairy tale of endless growth began to spin faster and faster, oil-fueled. More goods could be produced quicker and cheaper, for ever-new groups of consumers. And when population slowed down in the "first world," the rest of the world came into focus to be turned into consumers too. The legendary French politologist André Gorz recognized that "the mainspring of growth is this generalized forward flight, stimulated by a deliberate sustained system of inequalities."[4] Inequalities because, as soon as the majority of people can afford a product or service which once was the exclusive privilege of an elite (like foreign travel, a car, a computer), it is thereby devalued. The industry will immediately present the next upgrade or new product to aspire to, "endlessly creating scarcity in order to re-create inequality and hierarchy." In this game of carrot-dangling,[A] capitalism creates more unfulfilled needs than it satisfies: "the growth-rate of frustration largely extends that of production."[5]

Take the social ideology of the motorcar. It was an absolute privilege in the 1960s, and has been seen as an icon of freedom ever since. But long since there are too many and air pollution is a big problem.[B] But while we sit in the traffic jam, we can enjoy the next big premium product: the mobile cellular phone. The same story all over again: freedom and prestige for the first owners > once everybody had a mobile we needed smartphones to "be better than thy neighbor" > then version 4, then 5, then 20. As with cars, we're now competing about traffic speed on the data highways.

The next craze on the horizon is the smart home where almost every item shall be digitized. Your heating, fridge, kettle, etc. all shall be interconnected, with a digital interface, maybe a voice like Alexa, and a straight data-line to share all your likes and habits with the corporates' marketing departments. Can you imagine how much electricity this multiplication of electronic services would need? Plus all the data communications from self-driving vehicles connected via the highly energy-demanding 5G system?

And then there's smart farming. In the USA, automated harvesting of lettuces and strawberries has long arrived. And in France, robots prune grapevines and do some weeding and cultivation. A farming robot can recognize up to 800 different kinds of weed, destroying them with "precision targeting."[6] Already now, livestock is increasingly "cared for" entirely by automatons; basically, the animals spend their whole lives inside a super-barn-sized machine.[7] The spooky rural landscapes of tomorrow shall be human-free and populated by drones, robots, GPS-controlled tractors and harvesters with big-data connection.[8] Farming is a major target of the Digital Revolution. What the blue-eyed prophets won't mention is that, fundamentally, robots can only harvest crops suited to be machine-harvested in the first place, which calls for more streamlining and conformity, even bigger monocultures, and ideally the "consistency" of GMO clones. And the new electronic farms will cost a fortune, but that's what loans are for. Public subsidies can help the farmers to go deeper into debt, the banks can earn more than ever, so can the global food syndicates, and GDP will be *growing!*

But take courage! This brave new world will never fully descend upon us because of a fundamental energy problem. The carbon footprint of the Internet is already immense. Worldwide, the electricity demand on Internet-connected devices, emails, surveillance cameras, and mobile devices is increasing 20 percent a year. In 2010, the Internet released around 300Mt (million metric tons) of CO_2 worldwide, more than half of the fossil fuels burned in the UK.[9] In 2015 it swallowed about 3-5 percent of the world's electricity.[10] In 2016, the server farms in the USA needed the equivalent of about 8 big nuclear reactors, or twice the output of all the nation's solar panels[11]—except that they are largely powered by five old and dirty coal power plants in Virginia.[C] Greenpeace calculates that so far only 20 percent of the world's Internet electricity is renewable, with 80 percent of the power still coming from fossil fuels.[12]

The same situation of electricity demand and wide-spread use of the oldest, dirtiest coal power plants applies to bitcoin. The so-called "mining" of the cryptocurrency began in 2008 and ten years later its annual worldwide energy consumption had grown to 31.29TWh (terawatt hours = trillion watt hours),[13] or in less conservative estimates even twice that: about 7 gigawatts (at any given time), equivalent to the energy use of Switzerland.[14] A 2018 study published in *Nature Climate Change*[15] warned that, on its current growth rate, bitcoin alone could produce enough CO_2 emissions to push global heating above 2°C as early as 2033, but this calculation has been strongly contested (of course). More conservative estimates about bitcoin's energy consumption and carbon footprint range from 17[16] to 22.9 million metric tons of CO_2.[17] The point is that bitcoin is not climate-neutral, as claimed by some, by any stretch of the imagination.

And all these numbers are still not including smart homes, smart cars, smart farming, and the total digitization that is scheduled to replace up to half of all jobs.

By the way, charging your smartphone takes about 1KWh per year, that's not much. But then, a single snowflake is not aware of the avalanche it's part of.

As for Cheap Oil, it is not lasting forever, so something has to change anyway. There seems to be a wide agreement that "Peak Oil" occurred in 2005. This is the term for the maximum oil extraction. The maximum pumping rate begins to decrease once half of a deposit is emptied. From then on, production becomes slower and more difficult. There might still be oil for decades to come, but its extraction becomes more and more expensive, *until the point when it takes more energy to get it than what the yield would provide*. We are already in the phase of desperate deep sea drilling, tar sands, and fracking.

The economist Jeremy Rifkin, advisor to the European Union since 2000, and to Germany, and China, expects the fossil fuel industry as we know it to collapse by 2028, plus/minus five years.[18] This might have seemed far-fetched to outsiders, but only until the global oil price crashed—even below zero—during the corona lockdown in April 2020.[19] Over the corona summer, bankruptcies in the industry began to mount. In the USA the legal battles about who is going to plug up abandoned wells have begun. Oil companies plan to pay their executives, to the last minute, millions of dollars as boni and simply transfer the cleanup costs to the state, effectively leaving the mess to the public. Since uncapped wells leak methane and oil, here is *another historic high-priority job for the climate emergency course of action: to properly cap and seal the wells of the fossil fuel industry*. The independent climate-focused financial think tank Carbon Tracker has estimated the costs to plug the 2.6 million documented onshore wells in the US at $280 billion (not including an estimated 1.2 million undocumented wells).[20] And this is only the USA.

Just before the corona shutdowns, solar and wind energy had become cheaper than fossil fuels. And it was a victory for British campaigners when in March 2020 the UK government ended a block on onshore wind turbines and solar power from 2015.[21] Worldwide, wind and solar plants are already cheaper than electricity from about 60 percent of coal plants.[D] In China and South Korea, renewables prices began to undercut coal energy in 2020. This is expected to happen in Japan in the mid-20s for solar and by 2028 for wind.[22]

While the share of renewables in the energy mix has been constantly rising, the overall demand on energy has nevertheless outstripped the spread of renewables. The energy output of renewables might have doubled from 1973 to 2018, but the world total energy demand grew in the same period from 6Mtoe (megatons oil equivalent) to almost 14.3Mtoe. Hence the renewable share even shrank. The decrease of fossil fuels looks good in the statistics (down 5.4 percent) but their usage more than doubled (from 5,300 to 11,600Mtoe), and so did their GHG footprint.[23] In other words: *overall economic "growth" still by far outweighs the total performance of the renewables*.[E]

Furthermore, large-scale renewable energy plants, although being called "sustainable", are not free from serious unwelcome ecological impacts. For example, the foundations and towers of big wind turbines require massive amounts of concrete. And the compounds in photovoltaic solar panels are anything but green in their production: silicon requires quartzite or sand, gallium requires bauxite, tellurium requires copper or lead, and cadmium requires zinc—all of which lead to landscape destruction via mining and industrial production plants. At least, most parts of solar panels can be recycled. But another factor is the storage of electricity in batteries, and batteries are the worst; the cadmium and lithium (see "Back to Smoke & Mirrors" in Chapter 14) that make them function are still being mined in ecocidal ways and with appalling work conditions, and often children's labor.

Another major disappointment, or rather disillusionment, is "**biomass**". The UK is making reasonable progress in phasing out coal which is the dirtiest fossil fuel in terms of GHG emissions. A rise of renewable energy enables Britain to phase out coal rapidly. From 2016 to 2019, the share of coal in the British power mix fell drastically from almost a quarter to only 2.1 percent. And from 2012 to 2019, carbon emissions from the UK electricity grid have declined by more than two-thirds.[F]

So far so good. A substantial part of this success is the replacement of coal-generated energy with the output of biomass facilities. The original conception of "biomass" as a form of renewable energy is to burn wood chips, straw and other organic waste from agriculture and forestry. Essentially a good idea because these waste materials would decompose and release carbon dioxide over a number of years anyway, so they might as well serve energy production and thereby reduce fossil fuel usage. And this original idea is still the base for bold greenwashing by governments and the industry alike. In reality, biomass fuel has long shifted from gathering scattered organic materials to be purpose-grown in large-scale monocultures either of "energy maize" (which are responsible for the worst rates of soil erosion ever seen in countries like the UK) or of softwood trees.

Based on the false claim that burning wood for electricity is carbon-neutral, many European governments, last not least the UK, are funding such practices with large subsidies for biomass facilities. But these practices are not aligned with current science on reducing carbon pollution. In 2018 alone, UK's largest power provider, Drax, imported over 4.4 million tons of wood pellets from the southeastern US (here we go again, insane trade: shipping wooden fuel with fleets running on crude oil). Over 80,300 acres (32,500 hectares) of US forests were harvested for these exports to Britain. And it gets worse. Although the industry portrays itself as using "best practices" for biomass sourcing, mainly trees from thinning pine plantations, it turns out that Drax's supplier, Enviva Pellets, the largest wood pellet manufacturer in the world, also clearcuts hardwood forests in previously intact ecosystems, including whole trees and other large-diameter wood. Mature trees are the most carbon-intensive sources of biomass.[24]

Canada jumped the bandwagon and has become the world's second-largest exporter of wood pellets. It exports to the UK and to Japan. Canada and British Columbia are generously subsidizing the development of wood pellet exports and praise them as a climate solution. Here too, old-growth forests are not safe from extraction, even mature redcedars (*Thuja plicata*) from BC's unique rainforest have being trucked into Pacific BioEnergy's pellet plant. What's particularly painful about this subsidized "climate solution": "Wood-burning biomass is the worst carbon polluter of all, worse than coal, worse than oil, and worse than gas," said Mary Booth, Director of the Partnership for Policy Integrity. "Forests need to be protected as part of a global solution to the climate crisis. There is no path to achieving our climate goals if we continue to burn forests instead of conserving and expanding them."[25]

Already in late 2017, an open letter from scientists to EU parliamentarians pointed out that for some years a regrowing forest absorbs less carbon than if the old forest were left untouched. The reabsorption of carbon will take decades to centuries, time we don't have now. "At a critical moment when countries need to be 'buying time'," the letter states, "this approach amounts to selling the world's limited time."[26] A month later, another 800 scientists wrote a letter to the EU Parliament to point out this madness, criticizing that the current directive lets countries, power plants and factories claim credit toward "renewable energy targets" for deliberately cutting down trees to burn them for energy. "Biomass" needs to be restricted to residues and wastes. "Burning wood is inefficient and therefore emits far more carbon than burning fossil fuels for each kilowatt hour of electricity produced."[27]

Meanwhile, Canada itself gains 60 percent of its domestic energy needs from hydro power. 900 large-scale dams turn Canada into a hydro superpower. But while hydropower is certainly clean and renewable, it cannot be considered "green", as even the US Environmental Protection Agency (EPA) knows. First, there is the massive displacement of human settlements, habitat loss and destruction of entire local ecosystems. Then there is the massive energy cost and carbon footprint (concrete) in building the dams, and there is the impact of the carbon and methane released by vegetation rotting in flooded reservoirs. What's more, naturally occurring mercury is unlocked from the flooded soils and vegetation and enters the water, where it is transformed by bacteria into methylmercury, a neurotoxin that ends up in the food chain, accumulates in and poisons fish, waterbirds, seals, and humans.[28]

In summary, *industrialized* "green" energy is largely a delusion, not meant to protect the Earth but to protect our deeply flawed extractive economy. The narrative that simply decarbonizing society and replacing fossil fuels with renewable energies will save us all has long been the false prophecy of the very neoliberal globalism which has brought the ecological crisis upon us in the first place. Tragically, many environmental organizations and green parties have bought into the idea that with renewable energy everything will be fine and the extractive growth economy can "sustainably" keep expanding forever. The emperor's new clothes.

It has to be said that renewables will never be able to provide, in an ecologically sound way, for the energy needs of a world population in overshoot mode, let alone one hailing perpetual material growth. In fact, no means of energy production can, because the Earth is a finite planet. The neoliberal growth ethic is an "orchestrated illusion", a socially constructed cultural meme which we can, and must, liberate ourselves from. The only truly "sustainable" way forward is to abandon the economic growth trajectory (Degrowth, see box) and decentralize our culture, that is, *to produce as much energy and food as possible locally.*

Less is enough: Degrowth

> "Degrowth is the purposeful contraction of the economy in a controlled manner until it is in a sustainable steady state in which there is a more or less equitable distribution of the products of the economy within the means of nature."
> – William E. Rees[29]

Perpetual material growth is not even theoretically possible. In the physical world, all material and energy exchanges are subject to the second law of thermodynamics. It means that any change in an isolated system increases the system's *entropy* (a measure of disorder or randomness). Each exchange increases the degree of disorder: electrical charges wear out, chemical concentrations disperse, temperature or other gradients level out and disappear. Eventually, a permanent state with no energy for change is all that remains. "Maximum entropy" denotes that no change is possible.

Earth's ecosphere, however, with its thousands of complex ecosystems, myriads of life forms and countless interactions, "evolves and maintains itself in a dynamic steady-state by dissipating an extra-terrestrial source" of energy, sunlight.[30] Ecosystems self-produce, slowly accumulate biomass (such as plant material), and one being's or system's waste is another's thriving ground (see "Waste management", Chapter 4). Life itself seems to contradict the second law of thermodynamics by increasing order and decreasing entropy.

For the longest time, the human subsystem has co-inhabited this framework. Humanity's expansion across the continents and into ever-new territories did cause habitat destruction and ecocide on local and regional scales, but never became a threat to the ecosphere on the planetary level. Until the use of fossil fuels...

Modern humans, in the biological sense, have existed for about 200,000 years, and it took 99.9 percent of this time for the human population to reach one billion. But when our species committed to the burning of fossil fuels as a primary energy source, it only took a couple of centuries, or 1/1000th of the time span of our existence, for the human population to multiply from one to almost eight billion. The surge in energy, food and wealth allowed us to exponentially grow, consume all the planet's resources, and spill (toxic, largely unrecyclable) waste everywhere.

The human "outbreak" as such is perfectly natural; all species have a predisposition to expand swiftly when conditions are excellent. Think of spreading mold on old bread, or an algae explosion in anthropogenically over-fertilized waters. In ecology, this stage of a species' cycle—thriving on the temporary abundance of resources—is called the *plague phase*. It naturally lasts until negative feedback—food/energy shortages, disease, predation (depending on species and circumstances)—kicks in and reduces the population and its consumption levels again to degrees within the carrying capacity of the ecosystem.

Currently, we are financing economic growth by "liquidating the biophysical systems upon which humanity ultimately depends." (Rees)[31] We wreak havoc on the ecosphere. Climate disruption and loss of biodiversity are sure signs of increasing entropy. The world is dissipating, and the "human enterprise will almost certainly be forced to contract by energy/food/etc. shortages or foundering life support systems." Contraction is coming, one way or another. But we still have a choice whether to dismount and shrink in a graceful, orderly manner, or simply descend into pandemonium. "Chaotic collapse is probable and the usual outcome for societies whose leaders ignore evidential warning signs or are too corrupt or incompetent to act accordingly," warns Rees. The other option is degrowth, a controlled descent that respects the planetary boundaries and eliminates overshoot. This requires collective solutions and unprecedented international cooperation.

The concept of degrowth is gaining momentum. In September 2018, over 200 academics sent an open letter to the European Union and its member states to plan for a post-growth future in which human and ecological well-being is prioritized over GDP. The letter was published by *The Guardian*,[32] and over 15 European newspapers. A subsequent petition reached over 90,000 signatures.[33] The discussion about degrowth has just begun...

Footnotes

A. In the USA, for example, overall wealth has strongly increased since the 1970s, but satisfaction and happiness have not. Check the Happy Planet Index: http://happyplanetindex.

org/countries

An example of carrot-dangling: smartphones. In 2018 there were 2.5 billion smartphone owners in the world (36 percent of the world population), 210 million more than in the previous year. But the number of phones sold in that period was 1.54 billion. That means, 1.3 billion, or half of all smartphone owners, have replaced their phone with a new one without real need.[34]

B. Running a car demands more than we tend to admit. Depending on our wages and size of car, if we count not only the travel time but *all* car time, including the hours we have to work in order to buy, register, insure and maintain it, the total time we invest into traveling by car can easily amount to the same time per mile as if we had been walking. Except that we cannot walk anymore, because suburbs and the whole infrastructure have been built around the car. "To make room for the cars," says Gorz, "distances have increased. People live far from their work, far from school, far from the supermarket." Cars have created more distance than they have overcome. And "since cars have killed the city, we need faster cars to escape on superhighways to suburbs that are even farther away. What an impeccable circular argument: give us more cars so that we can escape the destruction caused by cars."[35] (The same insane argument is used to preach economic growth.)

What promised to be a wonderful independence has become a radical dependency. Way back, a rider could simply graze her or his horse, a cart owner could mend a broken wheel, but modern drivers are totally dependent on a maintenance system and—the oil industry which fires up a customer base of global proportions (which is intended, and automatically increases with population growth).

C. Loudoun County in Virginia is the home of data centers used by about 3,000 tech companies; about 70 percent of the world's online traffic is reckoned to pass through Loudoun County. Of course, Silicon Valley is all nice and clean; the headquarters of Apple, Google & Co. put their environmental conscience into action and use renewables. But a single new Apple data center planned in Co Galway, Ireland, is expected to eventually need over 8 percent of the national energy capacity; when the national grid will be low on wind power, there will be 144 large diesel generators as backup.[36]

D. Carbon Tracker reported in March 2020 that coal developers and investors risk wasting more than $600 billion because it was already cheaper to generate electricity from new renewables than from new coal plants.[37] On the other hand, policymakers and industrialists alike hang on to the old calculations that over 30 years, money invested in fossil fuels generates almost six times the amount of energy than wind turbines, and eight times the amount of solar arrays.[38] That is probably why so little is happening on the big scale.

E. Solar and wind now produce about 1850TWh of energy worldwide. But the global increase in energy demand in 2018 (2.9 percent or 938TWh) was 60 percent higher than the total output of all existing solar photovoltaic installations (585TWh).[39]

F. From 507 to 161g of CO_2 per kilowatt-hour of energy.[40] In the corona lockdown month of April 2020, the emissions even fell to 143g/KWh.[41]

Chapter 14

Climate Disruption

Climate change is the defining challenge of our time. We are currently way off track to meeting either the 1.5C or 2C targets that the Paris agreement calls for. [...] Time is fast running out for us to avert the worst impacts of climate disruption and protect our societies. [...] We need [...] to ensure a safer, more prosperous and sustainable future for all people on a healthy planet.
– António Guterres, UN Secretary General, 2020[1]

In the global average, the six years up to and including 2020 were the hottest six years ever recorded.[2] Weather extremes ("freak weather") such as severe cyclones, hurricanes, floods, and droughts, which used to occur perhaps once in a decade, now happen every year, sometimes every month or week. Due to global heating, heatwaves in Europe are now five times more likely than they used to be, and the scorching summer of 2019 led to 20,000 emergency hospital admissions and 1,462 premature deaths in France alone. The USA saw heavy rains (from July 2018 to June 2019) being the highest on record, and total economic losses at an estimated $20bn.[3]

In the summer of 2007, Arctic Sea ice cover reached an all-time low. In summer 2012 the ice in the Arctic Sea covered only 46 percent of that in 1980, and almost the entire surface of the Greenland ice sheet began to melt. Only two years later it was down to 40 percent.[4] In February 2018, the Arctic experienced its fourth winter heatwave, with temperatures rising above freezing four years in a row. The 2018 heatwave reached a temperature of 6° degrees Celsius (43 degrees Fahrenheit) at Greenland's northernmost observatory, just 440 miles from the North Pole.[5] The thawing of permafrost soil continues, greenhouse gases (particularly methane) from the frozen ground will accelerate the climate crisis seriously.[6]

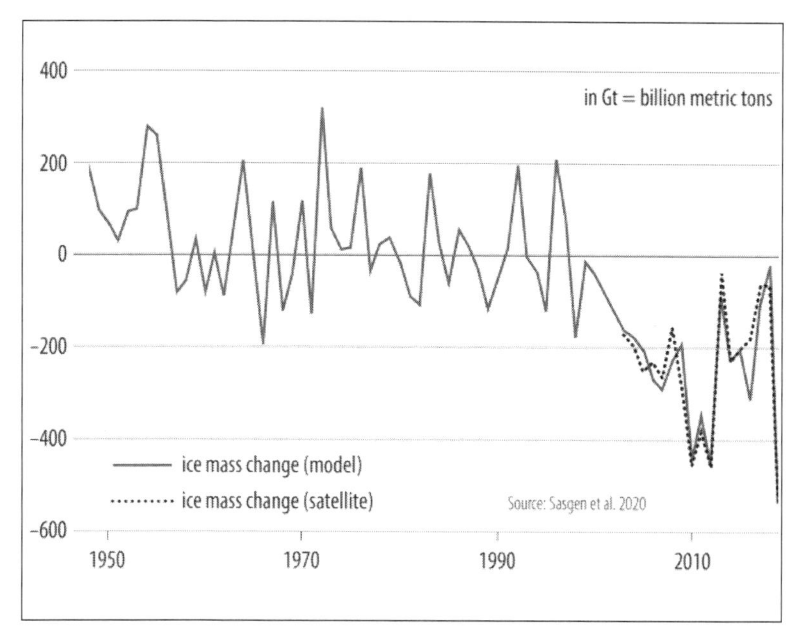

Figure 13. Changes in the Greenland ice sheet (in gigatons)

The World Glacier Monitoring Service reported that 2018-19 was the 32nd year in a row in which more ice was lost than gained.[7] The polar ice caps are now melting five times faster than in the 1990s.[8] The Greenland ice sheet reached a new record low by losing 532Gt during 2019, equivalent to one million tons per minute.[9] And since it took about 30 years for the ice caps to react to global heating, a further three decades of melting is inevitable now, even if human-made emissions were halted today. With the warming of the Arctic circle, the permafrost soils have begun to thaw. They contain about three times as much carbon as the atmosphere currently does; if only a fraction of this is processed by microorganisms and released—and it will if global heating nears and exceeds 1.5 degrees Celsius—we could not keep the world temperature from rising only 2 degrees Celsius; it will go up 4 or 5 degrees Celsius.[10] Melting glaciers also destabilize geological formations,[A] and melted ice dilutes the saltwater of the Arctic Sea which weakens the Gulf Stream (see "Tipping points" at the end of this chapter).

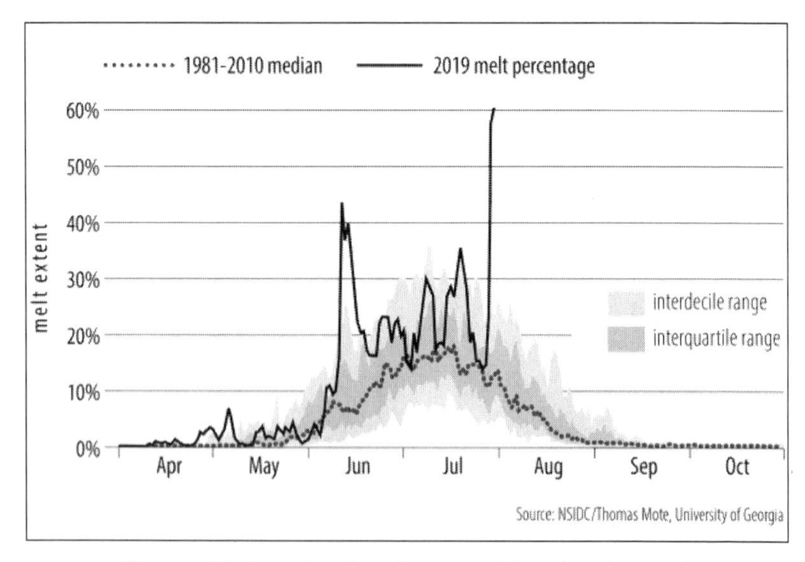

Figure 14. Greenland melt extent 2019 (in gigatons)

The oceans too had never been as warm as in 2019, with at least 84 percent of the seas experiencing one or more marine heatwaves (compare "Mass Mortality Events", Chapter 9).[11] Because warmer waters mean more evaporation this leads to severe disruptions in the planet's water management aka climate: drastic increases in air humidity which result in more (sudden and violent) cyclones, rainstorms, floods, etc. And there is a direct danger to humans: high humidity counteracts the efficiency of sweating. In high temperatures, the well-being of the human body relies on sweating, because sweat cools the skin as it evaporates. But when air temperature reaches 95 Fahrenheit (35°C) in combination with near-maximum air humidity (so-called wet-bulb temperature), any human activity becomes near-impossible, and survival increasingly jeopardized.

Already now, "even in the abundantly air-conditioned U.S., heat currently kills more people than cold, floods or hurricanes" (*Scientific American*).[12] We are propelling ourselves out of the benevolent climate niche which has supported us for the last 6,000 years. At a global average temperature rise of 2.5 degrees Celsius above the preindustrial level, wet-bulb temperatures higher than the 35-degree-Celsius (95 Fahrenheit) limit will become a regular occurrence in many regions of the world. For every additional 1-degree-Celsius rise in global temperature about one billion people will either be displaced or forced to endure insufferable heat.[13]

Australia suffered a terrible heatwave which led to Christmas temperatures as much as 12°C above the December average; parts of Victoria reached temperatures of 104-115 Fahrenheit (40-46°C).[14] That summer sparked the Australian youth climate movement, but was even topped by temperatures one year later which resulted in wide-spread and long-lasting catastrophic bushfires of unprecedented scale and vigor. The fires caused mass evacuations (more than 100,000 residents), contributed largely to atmospheric CO_2 levels, destroyed millions of acres of land, and killed about one billion mammals, birds and reptiles.[15] Humans caught up in the ordeal described their experiences as apocalyptic.[16]

In Brazil, the 2019 fires in the Amazon rainforest rose 30 percent over those in the previous year;[17] and the 2020 fires rose another 61 percent.[18] And in 2020, Argentina, Bolivia, and Paraguay were ablaze too. Because of widespread wildfires, fanned by strong winds and temperatures above 113 Fahrenheit (45°C), Bolivia declared a state of emergency in mid-September; and with more than 12,000 outbreaks, neighboring Paraguay followed suit on 1 October.[19]

In California, wildfires burned about twice as long in 2018 as they did in 1990, and the average number of acres burned per fire had doubled.[20] In 2020, the extent of wildfires reached a new order of magnitude: while close to 260,000 acres (105,000 hectares) had been destroyed in the year prior, in 2020 it was well over 4.3 million acres (1.76 million hectares). That's a rise of sixteen times over. On top of that, the first "gigafire" in modern history merged from a number of already serious ones, to range over 1 million acres.[21] And the phenomenon of wildfires is migrating north. By September 2020, fires had burned more than 900,000 acres in Oregon, and half a million people—about 10 percent of the state's population—had to evacuate. In Washington state more than half a million acres burnt.[22]

The Arctic has been heating up more than twice as fast as the rest of the world, resulting in ever more severe heatwaves and even wildfires, especially in Siberia. On 20 June 2020, the Russian Meteorological Service measured the highest temperature ever recorded beyond the Arctic circle: 100 Fahrenheit (38°C)—in a town more than 400 miles farther north than Anchorage, Alaska. During the first half of 2020, wildfires ravaged over 7,900 square miles (20,460 square kilometers), an increase of 16 percent over the previous year.[B] These fires released 56Mt of CO_2.[23] Due to drought and heat, trees in areas not affected by fires often become a breeding ground for large swarms of Siberian silk moths (*Dendrolimus sibiricus*) that feed on the trees and make them more prone to future fires. But not all of the fires burn forests: about half of them occur on peat soils. The peat soils of the world contain hundreds of gigatons of carbon.[24] The warming of the Arctic has long begun to thaw permafrost soils which in turn will release methane from the breakdown of ancient plant materials.[25] And higher temperatures increase positive feedback loops by melting ice that used to reflect solar energy (decreasing the planet's *albedo*, see Chapter 5).

Reasons for wildfires

While global heating with its drier weather, droughts, and heat waves increases the odds that wildfires become bigger, longer-lasting, and hotter, the initial sparks have manifold reasons.

In **Central and South America**, most fires are initiated to clear land for cattle grazing and soy production, sometimes for real estate deals, but in every case it is powerful businesses that benefit a very few people at the expense of the rest of society, and of nature. In **Asia**, particularly in Indonesia, it's about big money and oil palm plantations. In **Africa**, fires cover even much larger areas than in the Amazon, but the fire seasons on these two continents cannot be compared. In the savannahs of Angola and Mozambique, annual fires are part of the ecosystem and mean renewal. In the Democratic Republic of Congo, many of the fires are located outside the Congo Basin's sensitive rainforest areas. Mozambique and Madagascar do lose (far too much) forest to small-scale slash-and-burn farming, but their governments try to change that.[26]

The bushfires in **Australia** are dominantly caused by changes in the weather patterns: about half of them is sparked by lightning. Ignition by humans is rather by accident than arson. Records going back to 1950 show that Australia is steadily becoming hotter and drier,[27] connected to ozone depletion in the stratosphere and climatic changes in Antarctica and the Southern Pacific.[28] Australia's average temperatures are now 1.52°C above the 1961-1990 average, and extreme heat and dryness leading to fires are considered four times more likely because of human-caused climate disruption.[29] Especially parts of southern and eastern Australia develop more dangerous conditions during summer and an earlier start to the fire season. Prof. Nerilie Abram, a climate scientist at the Australian National University, adds: "It's worrying that we are talking about this as a new normal, because we are actually on an upward trajectory. Currently the pledges in the Paris agreement are not enough to limit us to 1.5°C—we are looking more like 3°C."[30]

In **California** and the US West Coast, the reasons are complex in a different way. In pre-history (pre-1800), vast areas in California burnt annually. For thousands of years, hundreds of Indigenous tribes across California used small, controlled fires to remove fire-fueling vegetation, renew the soil, and prevent bigger wildfires.[31] They worked in partnership with nature. After all, fire is an essential part of the Californian ecosystem, redwood trees (*Sequoia*) have developed a fire-resistant bark, their cones have evolved to open with the heat of a fire, releasing seeds that germinate best in freshly burned, fertile forest floor.

But starting in the 1880s, European settlers outlawed the ecological and cultural practice, and began fining Indigenous people for burning their own lands. In their limited understanding wildfires were a "moral and mortal enemy of forest" because they damage agricultural land and valuable timber. As a result, brushwood, leaf litter, and dead trees have been building up for decades, turning many areas across the country into virtual tinderboxes. Additionally, the climate crisis accounts for more than half of the increase in fuel aridity in western US forests.[32] Even the giant sequoias, weakened by warmer, drier weather and severe droughts, are less and less able to withstand progressively hotter and longer-lasting fires. However, 95 percent of all wildfires in California are started by human activity. According to the California Department of Forestry and Fire Protection (CAL FIRE) "this could be campfires, illegal burns, improperly maintained vehicles, and the worst cause, arson."[33] Another considerable fire hazard in CA is the outdated electric grid.

Don Hankins, a pyrogeographer and Plains Miwok fire expert at California State University, Chico, notes that the overall apocalyptic outlook does not mean that California is unlivable: "In most places in California where these fires are happening, Indigenous people have lived in these places for thousands of years, with fire." And across the west, policymakers will have to restore Indigenous land stewardship and work with Native American fire practitioners to restore the preventive practice of burning land in a controlled way (as much as global heating will allow).

The main reason for **Arctic** wildfires is that temperatures in the polar regions rise disproportionally faster than the global average. This is where the climate crisis reveals itself most unmistakably. An international study showed that the prolonged heat in Siberia during the first half of 2020 was made at least 600 times more likely as a result of human-induced global heating.[34]

So are we stopping the causes for this catastrophic deterioration? Global heating and the resulting climate disruption are due to anthropogenic (human-made) greenhouse gas (GHG) emissions. This has been known for decades, but global GHG emissions keep increasing, carbon emissions jumped to an all-time high in 2018,[35] and rose to a new record in 2019.[36]

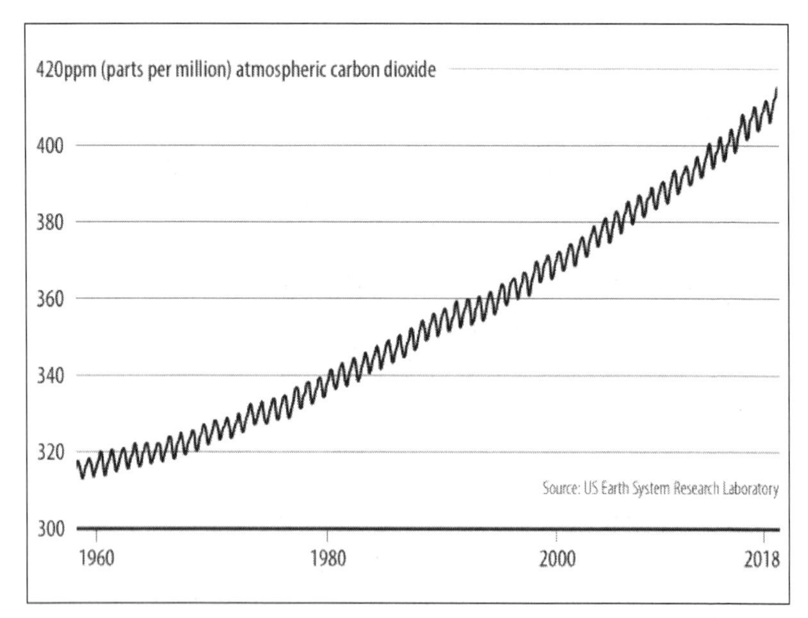

Figure 15. Atmospheric CO_2 measurements from Mauna Loa, Hawaii
The steady rhythmical variation represents the fact that the vegetation in the Northern hemisphere absorbs atmospheric carbon in its growth phase in spring and releases carbon again each autumn.
(This figure is a close-up of the vertical carbon line in Figure 17.)

Although greenhouse gas emissions dropped about 6 percent in 2020 due to travel bans and economic slowdowns in response to the COVID-19 crisis, this improvement was only temporary. Global heating was not on pause, it always keeps accelerating.[37] Diverting it requires measures as drastic as those implemented for COVID-19. As the deep green ecologist Sandy Irvine warned: pandemics like *"Covid-19 will not create an uninhabitable Earth. Unchecked ecological meltdown will."*[38]

If you think that speaking of a global ecological meltdown is over-dramatic, consider this: one-fifth of the world's countries are already at risk of their ecosystems collapsing because of habitat destruction and ecocide. According to an analysis by a Swiss insurance firm more than half of global GDP—$42tn

(£32tn)—depends on a healthy biodiversity, but the risk of tipping points is growing fast. Australia, Israel, and South Africa are most at risk to lose their "ecosystem services", with India, Spain, Belgium, Pakistan, and Nigeria also being marked as highly fragile. Among the G20 leading economies, the USA is the 9th and the UK the 16th most endangered nation.[39] Brazil and Indonesia should really start protecting their remaining wild nature rather than closing in on it.

And the climate breakdown is not theory either. It has long arrived in measurements and observations, and direct, raw experience: in 2019 alone, about 24.9 million human beings across 140 countries became displaced by environmental and climatic "changes", and 5.1 million of these remained homeless climate refugees.[40] And this number will grow exponentially if business-as-usual is allowed to continue. So why is "climate denial" still so strong?

Smoke and mirrors: Strategies of disinformation

Corporate

A climate study marked "confidential" and presented to the board of directors of Shell Oil stated—in clear words and beyond doubt—that CO_2 emissions from the burning of fossil fuels would heat up the planet. Shell summarized that "the changes may be the greatest in recorded history," and listed implications such as the decrease of polar ice caps, rising sea levels (possibly as much as "five to six meters"), the disappearance of specific ecosystems, habitat destruction, increasing problems with food and freshwater accessibility, diminishing affluence in various parts of the world, and inevitable mass migration.[C] This internal Shell report is from—1986.[41]

It wasn't even the first one of its kind. An internal ExxonMobil report from 1982[42] predicted that global CO_2 levels would, from their preindustrial level, double to 560ppm (parts per million) by about 2060, and would push the planet's average temperatures up by about 2°C.[D] The right move would have been to inform governments and the public, but instead, "oil firms recognized that their products added CO_2 to the atmosphere, understood that this would lead to warming, and calculated the likely consequences. And then they chose to accept those risks on our behalf, at our expense, and without our knowledge," says Stanford climate activist Dr. Benjamin Franta, and asks, "Who has the right to foresee such damage and then choose to fulfil the prophecy?"[43]

Independent studies (from 1970, 1981, and 1988) have accurately predicted global heating for the past 50 years.[44]

In 1989 Shell and Exxon, together with BP and Chevron, founded the lobby organization Global Climate Coalition. The single goal was to use an annual multi-million-dollar budget to systematically cast doubt about climate science.[45] The organization was dissolved in 1998, and since then, Shell does not deny climate change any longer. But behind the scenes, funding against climate science was even increased:[E] an investigation from 2016 concluded that the entire fossil fuel industry spending on obstructive climate policy lobbying may be in the order of $500m annually.[46]

Shell's confidential report was first disclosed by the Dutch news organization *Inside Climate News* (ICN)[47] in April 2018. Exxon's classified papers were not intended for external distribution, either, but were leaked in 2015—ICN's string of articles was a finalist for the Pulitzer Prize in 2016.[48] Finally the world was beginning to wake up. Slowly, the "cast of characters gambling with the fate of humanity"[49] are being summoned to responsibility: calls for lawsuits, e.g. against Exxon under the Corrupt Organizations Act (RICO), have been increasing for a few years,[50] and support from scientists and the public is growing.[F]

Politics

The dark secrets about burning fossil fuels were also known outside Shell and Exxon from early on. Already in 1968, Gordon MacDonald, then science adviser to US President Johnson, published an essay wherein he predicted a near future in which carbon emissions could alter weather patterns and wreak famine, drought and economic collapse.[51] And in 1978, a group of elite scientists calculated that the doubling of the atmospheric CO_2 levels would be reached by 2035, and that global temperatures would increase by an average of 2 to 3°C. Their report was sent to practically all US institutions.[G] During the 1980s the USA came close more than once to halting climate change in its beginnings but repeatedly failed due to the resistance of Big Oil and a few hardliners in the Reagan and Bush administrations.[H] Additionally, some politicians too are masters of deception.[I]

Media

Mainstream media have crucially assisted the public misconception, not necessarily by intent. The simple dramaturgy of TV shows is to present discussions as "balanced" by finding a "skeptic" to argue with an independent climate scientist—it just makes good entertainment. But over twenty years this has "created a very unbalanced perception of reality. As a result, people believe scientists are still split about what's causing global heating, and therefore there is not nearly enough public support or motivation to solve the problem," explained environmental scientist Dana Nuccitelli already in 2013.[52] Slowly the message is sinking in, at least with younger generations.

The grooming to apathy has been going on for decades. It reached a new peak in the week of October 8, 2018. When the IPCC released its dire warning that humanity has but twelve years to get its act together, this was duly reflected at least in *The Washington Post*, *The New York Times*, and *The Guardian*. The latter published six to ten major climate articles each day (!), with headlines like "Our leaders are destroying our future". The NYT put out 24 big climate articles in the first two days after the announcement alone (and that's without the ones about the then ongoing hurricane), with appropriately drastic headlines such as "UN Report on Global Warming Carries Life-Or-Death Warning" and "Wake Up, World Leaders—The Alarm is Deafening". The *Post* was equally outspoken, for example, the article entitled "The world has just over a decade to get climate change under control, U.N. scientists say", started with these words: "The world stands on the brink of failure when it comes to holding global warming to moderate levels, and nations will need to take 'unprecedented' actions to cut their carbon emissions over the next decade."[53]

But this is not typical for the rest of the (media) world. The "news" media in most countries usually had *one* article about the IPCC's disastrous climate message, and not on top of the page either.[54] In Britain, the tabloids, read by millions of people, ignored climate altogether. *The Sun* headline on October 8 informed its readers that some ex-glamour girl had been arrested for drink-driving. Germany's biggest tabloid, *Bild,* stuck to a national disaster: "TV nun naked in traffic control". But even more prestigious publications increasingly mistake entertainment for news. Who are the press editors in charge to decide *beforehand* what the public will be interested in? Wherever you live, wherever you look: did your national or regional news channel/paper tell you, for example, that worldwide 207 environmental defenders were murdered in 2017, the highest number ever?[55] Most of them Indigenous people defending the land they love. Did your paper ever run an obituary to honor them?

As it is, the world populations are sleepwalking. When mainstream media finally revealed in autumn 2017 that all water for the foreseeable future is now being contaminated with microplastics, what happened? Nothing. Silence. At any time in history, if a stranger came to a village and poisoned the well, he would have been seriously prosecuted. But today, if you are a manager of a company that still keeps poisoning *all water* on Earth, instead of having to account for your deeds, you might be called to become a minister in the White House. After all, the legal concept of "limited liability" or limited responsibility for companies and chief executives has long mutated into *unlimited irresponsibility*.

In the words of Canadian author, activist, and political analyst Naomi Klein: "There is simply no way to square a belief system that vilifies collective action and venerates total market freedom with a problem that demands collective action on an unprecedented scale and a dramatic reining in of the market forces that created and are deepening the crisis."[56]

How can we expect politicians to install environmental measures that "burden" the corporates and the population if they want to get re-elected for the next term? Is it a surprise that each and every politician leaves the daunting stuff to whoever may succeed her or him? *The political system is not made for a task like the one at hand.* It's like the notion of the parliamentarian who says, "We know how to fix climate change, we just don't know how to win the election afterwards."

What can I do?

- Demand climate action from your national and local politicians, and more journalism and less distraction from the "news" media.

On the Internet, "climate skeptics" abound; their platforms might look "scientific" in their presentation and language but at second glance they always betray them. The most common characteristics are:

- Most of such websites haven't been updated since 2012; i.e. they thrived in the denial heydays during Earth's slightly cooler phase in climate oscillation (1998-2012, see below), and have been sticking around since.
- The "articles" merely claim "facts" but don't give sources. Often enough not even names of their authors. And if there are names they don't check out even on Google, and are not associated with a respectable institution (such as a university or the UN).
- Even arguments that appear logical often include statements that are simply wrong, or the writers inundate their websites with impressive but unreferenced diagrams (which they know their viewers won't verify anyway).

There is hardly a "skeptical's" argument that does not culminate in a core message that "climate change" is all but a hoax, or fear-mongering. To manipulate us. But it is never said just Who is supposed to be behind it and for what purpose. So the question is: What do the climate deniers want us to do? And it seems that they are saying: "Do nothing! Be silent. Leave things to run their course"—a course which happens to be controlled by Big Oil, Big Food, and Big Money. *You shall know them by their fruits.*

Misbeliefs and facts

This section clarifies the ten most widespread misconceptions about the climate crisis.

Misbelief no. 1: "Scientists themselves still don't agree whether global warming is man-made or not."

In fact, since the early 1990s, 97 percent of scientists who work in climatology agree that global heating is man-made.[57]

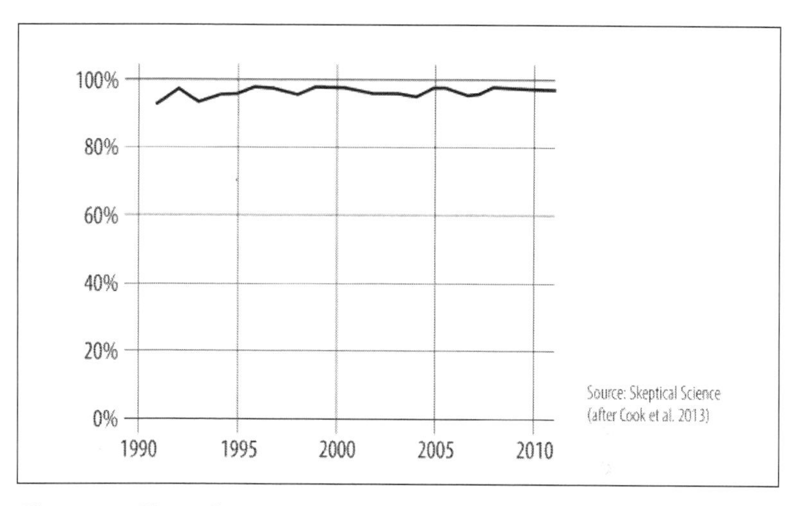

Figure 16. Since the 1990s, 97 percent of climate scientists agree that global heating is man-made

What can I do?

- Regarding the **97 percent consensus**, check The Consensus Project. theconsensusproject.com[58]
- Spread the word. Skeptical Science has a webpage where consensus graphics can be shared via social media or email: skepticalscience.com/graphics.php[59]

Misbelief no. 2: "Global warming has stopped! It's not getting warmer anymore."

This notion stems from the "cool" phase 1998-2012 when global mean temperatures—due to a number of variable climate parameters[J]—did not rise as much as expected in the 1990s.

The world is very different since 2013. Whatever the reason Gaia gave us a break, it is long over. Each of the years 2014 to 2016 became the hottest ever on record; 2017 fell just behind 2015 but was still the third-hottest ever. Then came 2020… The 5-year-mean value is back on track with the models, and the temperature curve is rising faster than ever before.

Misbelief no. 3: "Nothing is proven. It's all just faulty computer models."

There were faults in the early days, of course, models from the 1990s are different to those from the 2010s. Since then, climate science has made huge steps. But there is a big difference in the climate discussion now as opposed to pre-2013: it's not about modelling anymore, it's about observation. *Global heating is happening, it is measured by thermostats on land, on buoys in the world's oceans, and from satellites.* In 2018, Greenland's massive ice sheet melted at the fastest rate for at least 450 years. We regularly hear such drastic news about melting polar caps, shrinking glaciers, fish and birds migrating further north, fatal irregularities in monsoon seasons, and the whole palette of extreme weather scenarios. And all this is very much in line with the computer models.

Misbelief no. 4: "It is not even known if CO_2 has a climate effect."

Molecules the size of CO_2 and H_20 simply do catch heat rays that reflect from the Earth's surface and thereby they trap heat energy in the atmosphere, that's plain physics. Satellite measurements of infrared spectra over the last 40 years indicate the correlation with rising CO_2 levels.

Furthermore, climate history provides evidence: the Vostok ice core samples (together with analysis of tree rings and of coral reefs) give us information about the climate history of the last 420,000 years. Air bubbles encapsulated in the ice tell us the levels of CO_2 and other gases in the atmosphere over time. *Atmospheric CO_2 levels and mean global temperatures have always been closely correlated.* Atmospheric CO_2 levels fluctuate at 180-210ppm during Ice Ages and at 280-300ppm during interglacials. It is important to note that atmospheric CO_2 levels in this period never went below 180ppm or above 300ppm. The pre-Industrial level was not higher than 270ppm. But in 2013 we reached 400ppm, and in February 2018 already 407ppm. In

March 2019 411ppm and in June 2020 416.5ppm.[60] Visit www.co2.earth/daily-co2.

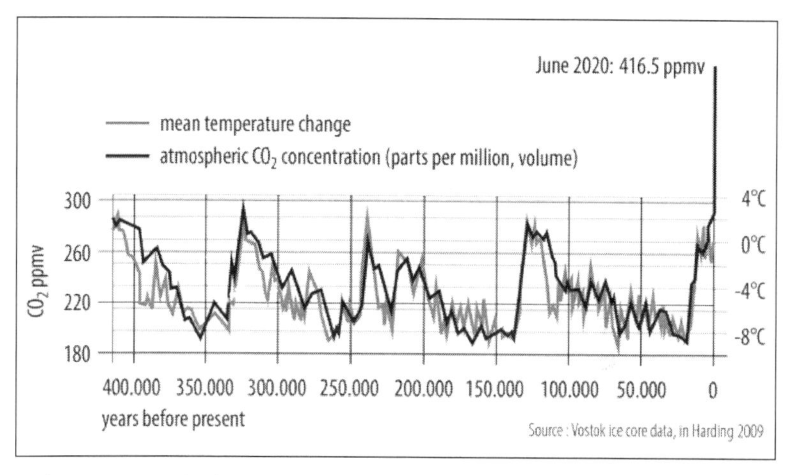

Figure 17. Variations of carbon dioxide levels and air temperature over the last 410,000 years

> The [CO_2 concentration] number is the closest thing to a real-world Doomsday Clock, and it's pushing us ever closer to midnight. Our ability to preserve civilization as we know it, avert the mass extinction of species, and leave a healthy planet to our children depend on us urgently stopping the clock.
> – John Sauven, head of Greenpeace UK, 2019[61]

Misbelief no. 5: "It is not even known if the extra CO_2 in the atmosphere is man-made."

Yes, it is! Carbon has only two stable, *naturally* occurring isotopes: 12C and 13C. Carbon from the burning of fossil fuels is a different isotope, 14C, a radionuclide which decays with a half-life of about 5,730 years (radiocarbon dating in archaeology is based on this). The amount of atmospheric 14C can be measured with spectroscopy. Carbon originating from burning mineral oil can also be identified in living plants. Secondly, carbon from burning fossil fuels also leads to a decrease in atmospheric oxygen O_2; this is being measured as well. Thirdly, national statistics monitor national carbon release: these figures comply with the measurements and models by climate scientists.

Misbelief no. 6: "Wasn't Greenland green before?"

Greenland ice has existed for at least 400,000 years, though perhaps some coastal areas where Vikings landed were temporarily "greener" than today.

Misbelief no. 7: "Didn't scientists in the 1970s talk of a new Ice Age?"

Yes, in about 50,000 years' time.[K]

Misbelief no. 8: "Is it not impudent to imagine that our insignificant race could have such a monstrous effect on an entire planet? Aren't volcanoes releasing much more greenhouse gases than we are?"

First, our race isn't insignificant anymore; we have ushered in the Anthropocene, a period when man-made fallout of

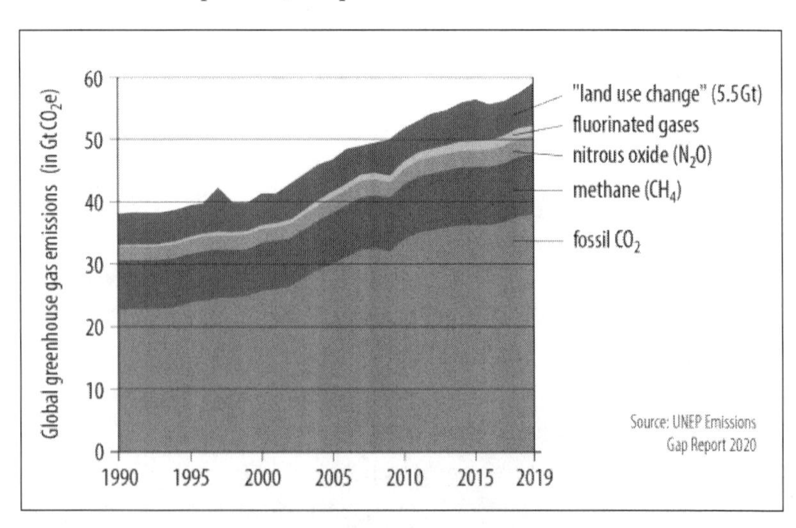

Figure 18. Global greenhouse gas emissions from all sources

In 2019, humanity emitted 59.1 billion metric tons of greenhouse gases (including shipping and "Land Use Change").

radioactive particles as well as microplastics have begun to leave a worldwide permanent mark in geological sediments. As for volcanoes, among the few currently active volcanoes, Mt. Kilauea on the main island of Hawaii has been active for the past 35 years. Mt. Kilauea spews about 7 million metric tons of CO_2 per year into the atmosphere. In 2019, annual human emissions reached over 38 billion metric tons, and for some reason, most media leave out the impact of "Land Use Change" (see next section) which brings the total up to 59.1 billion metric

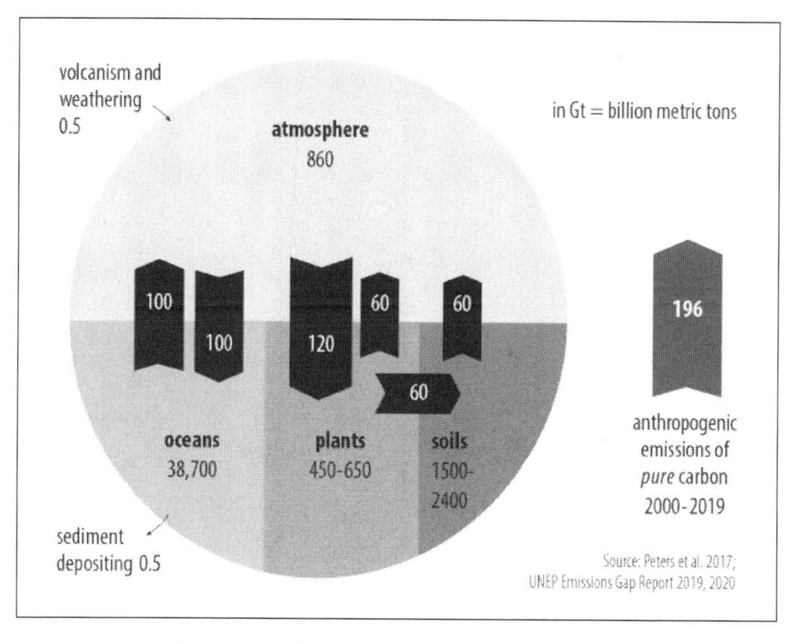

Figure 19. The natural global carbon cycle

It is obvious that in comparison to the massive amounts of carbon which the Gaian system turns over, the annual human impact does seem rather small. But the crux of the biscuit is that the release and the capture of carbon in the natural cycle is balanced in itself, while man-made carbon is *additional*. Secondly, anthropogenic emissions are *accumulating* because CO_2 is active for about 120 years. That's why we talk about humanity's "carbon budget" – we have almost used up the maximum of what we should ever have dared to emit.

tons. *Total annual anthropogenic greenhouse gas emissions are over 8,400 times higher than those of Mt. Kilauea.*

Misbelief no. 9: "Climate scientists are only in it for the money."

This is a projection of the venality of those who get paid by Big Oil to be "skeptics". At universities and independent research organizations, however, scientists don't earn nearly as much as they would in the fossil fuel industry. And that scientists actually do care (for the living world) you can see by their engagement in petitions, warnings to politicians, and the foundation of the Union of Concerned Scientists (ucsusa.org)[62] with more than 250,000 members. In December 2017 the Union renewed its *Warning to Humanity*, with signatures of more than 15,000 scientists.

Misbelief no. 10: "It's too late now anyway. It's not realistic to attempt cutting our emissions down to zero by 2050."

It's not realistic for the human species to follow the course of self-annihilation with open eyes and do nothing about it. Period.

The deep fear of big change harbored by politicians and laypersons alike is based on the *idea* that cutting down emissions, supposedly, equals less productivity equals less economic growth and thereby misery, agony, and death. However, it is non-action that will tumble us quicker into chaos than anything else. One key step is to *decouple* economic growth from employment and true prospering (well-being). Currently, the economy is based on the financial system of debt. It is the conventional economic paradigm that needs a drastic overhaul. Indeed, many alternatives are already in the pipeline, having been proposed by a number of top economists.[63]

Even the World Bank and an increasing number of fiscal agencies and economists say that *acting on climate change will cost only half* the other option which is doing business-as-usual and just runs *behind* each single destruction. In the end of the day, it is non-action that will increasingly destabilize each country. (Remember: coastal cities like New York and San Francisco are beginning to sue Big Oil for their decades of lies—it would have been much cheaper if we had started climate action back in the 80s.)

The carbon discussion

Monitoring emissions is necessary for reducing them. But then, this has been going on for decades, without triggering enough action anywhere near an adequate level. None of the ongoing discussions need to be repeated here, but we need to become clear about three things: *CO_2 accounts for less than two-thirds of climate-active greenhouse gases;* representation of transport and agriculture is usually heavily distorted; the influence of globalized freight shipping is not taken into account.

The hidden toll of agriculture and livestock

At first glance, transport appears to be the bigger source of GHG emissions, for example, about 29 percent in the USA, and 26 percent in the UK,[64] while agriculture is usually listed (by different organizations) at under 10 percent or a maximum of 14 percent, with about half of that being livestock. Hence, many people disregard it as a ridiculous idea that cattle should be significant climate factor at all. Transportation emissions are four or five times higher than those from livestock, or are they?

Such numbers may describe the situation in industrial countries, whose agricultural sector is *deceptively underrepresented in the GHG discussion* because of their high net imports of food and feed. Europe in particular has long outgrown its physical carrying capacity and for a large part lives on agricultural imports. *Not even half of the total agricultural and forestry output that Europe consumes or exports is produced on land located within Europe:* 53 percent of European land use is being virtually imported, mainly from Brazil, Russia, and China.[65] To tackle the global ecological crisis we cannot stop at national datasets. *In the planetary flux of substances there are no national borders.* Hence we need to look at global, not national data. Because the "rest" of the world drives fewer cars but grows more food and animal feed than the industrialized countries, the *global* balance is quite different to that of industrial nations.

The next major misconception is that in many statistics and graphs, the sector "Land Use Change" is attributed to forestry, which, however, requires comparatively little (fossil) energy. "Land Use Change" largely is a euphemism for the destruction of forest and other natural areas (usually by fire) and their subsequent conversion for the agricultural industry, mainly into pastures or monocultures for animal feed or palm oil. The consequent integration of "Land Use Change" into the agricultural sector of GHG emissions shifts the proportions significantly: agriculture 13.5 percent, and deforestation aka "Land Use Change" 18.2 percent of global GHG emissions.[66]

And the climate damage of burning forests is not just the carbon released by the fire, but the carbon which the trees won't

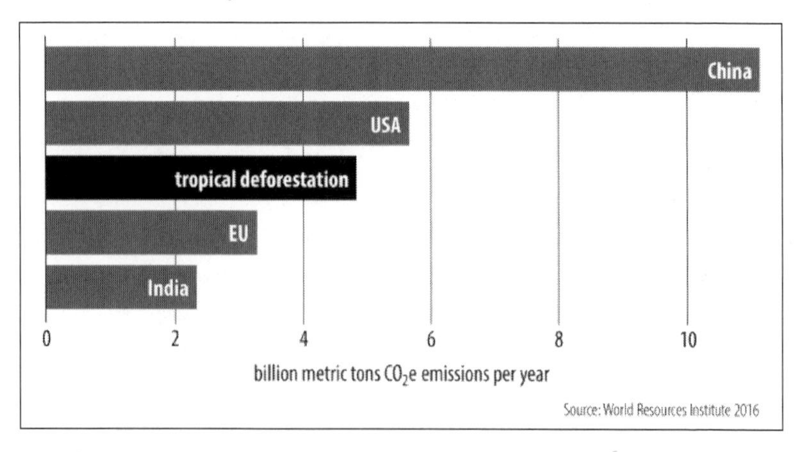

Figure 20. "Land Use Change": if tropical deforestation was a country, it would rank third in greenhouse gas emissions

absorb in the following years because they have been replaced by pasture, soy fields or palm oil plantations. Plus: Burning forests are a major source of nitrous oxide, a GHG around 300 times more potent than CO_2. If tropical rainforest loss was a country it would rank third in global GHG emissions, after China and the USA, in other words: tropical rainforest destruction produces more GHG emissions than the burning of fossil fuel in all of Europe.[67] This increases the responsibility for agriculture and livestock significantly.

The third widespread misconception is to focus on CO_2 alone. This greenwashes cows because cars undoubtedly emit more CO_2, and less methane. Only a quarter of the GHG emissions from livestock are CO_2.

Total global carbon dioxide emissions: about 36Gt = billion metric tons, in 2016[68] (37.2Gt in 2019).

But the *total GHG* emissions need to include the other gases that accelerate global heating, namely methane and nitrous oxide. They are calculated in carbon dioxide equivalent (CO_2 e):

total global greenhouse gas (GHG) emissions: about 50Gt CO_2e (2016).[69]

In this more honest scenario, the impact of transportation comes down to 16.2 percent (air, rail, ship & other 4.3 percent, road transport 11.9 percent) of all GHG emissions, while the impact of cows trebles:

total global GHG emissions from livestock: 7.1Gt of Co_2e = 14.2 percent (2016).

Cattle (beef and milk) represents about two-thirds (9.4 percent) of the livestock sector's emissions, and cars and vans constitute about two-thirds (8 percent) of road traffic emissions. Therefore, among livestock, cattle alone are worse for the climate than cars. And this is not the end of the calculation, there is more to consider:

What even the FAO doesn't seem to account for in this calculation is GHG from **soil degradation**. Through history, the intensely farmed soils of the world have lost about 133Gt of carbon, at least half of this global "soil carbon debt" is considered to have been caused by degradation of pasture lands. Some of this carbon lost through intensive agriculture is captured again by vegetation in other places, but a big part escapes into the atmosphere. This shows how essential the climate potential of soil management (and responsibility!) really is.

Additionally, the destruction of forests (mostly "Land Use Change" to crops and pasture) so far has released 15 percent of all greenhouse gas (GHG) emissions.[70] Today, the global food system is responsible for a third of all *annual* GHG emissions.[71] The global meat (and dairy) production requires 83 percent of all farmland, and causes 58 percent of all agricultural emissions (see Figure 21).[72] Among that, *cattle (beef and dairy) alone causes 12.4 percent of global GHG emissions* (dairy 0.75, beef 11.65 percent). That is more than all cars and vans (8 percent) in the world.

And last not least, there are shocking new findings about methane, which demand a further revaluation. Previously thought to be about 20 times more climate-active than CO_2, methane is now considered to be 86 times more damaging to the climate than the same amount of CO_2 over a period of twenty years. This tips the scale even further against breeding and eating burping ruminants such as cattle, sheep, or goats.

And if countries in the near future (in order to fulfil their Paris Agreement pledges for a world no hotter than 1.5°C above pre-industrial temperatures) cut GHG emissions across all the other sectors but leave livestock untouched, *livestock alone will be responsible for half of the world's greenhouse gas emissions by 2030.*[73]

This is by no means to say that the transport sector can be neglected! The situation is bad. But the global effects of livestock are even worse.

And there is the ethical/moral dimension. All human beings *could be* fed but with 83 percent of global farmland being dedicated to breed animals for meat, those countries who can afford to eat carnivorously express their self-importance while others suffer malnutrition. Racism doesn't just begin in the police force of rich countries. As Maneka Gandhi, then the Indian Minister for the Environment said already in 1990: "More grain is fed by the USA and USSR to livestock than is consumed by the people of the entire third world. Britain gives two-thirds of its home grown cereal to its livestock—that amount could satiate 250 million people each year."[74] She continued by pointing out that a tenth of Europe's meat and dairy products were being fed with fodder imports from India, thus depriving India of valuable water and land.

Livestock GHG emissions: 7.1Gt CO_2e (2016)[75]
Cattle (beef and milk) 65 percent, pig meat 9 percent, buffalo milk and meat 8 percent, chicken meat and eggs 8 percent, small ruminant milk and meat 6 percent. (The remaining emissions are sourced to other poultry species and non-edible products.)

Structurally, the 7.1Gt of livestock GHG emissions originate from:

- 45 percent feed production and processing (this includes "Land Use Change"),
- 39 percent enteric fermentation,
- 10 percent manure storage and processing,
- 6 percent processing and transportation of animal products.

Chemically, the 7.1Gt of livestock GHG emissions consist of:

- 44 percent methane CH_4,
- 29 percent nitrous oxide N_2O,
- 27 percent carbon dioxide CO_2.

Annual contributions of livestock supply chains to anthropogenic GHG emissions, in CO_2e (IPCC, 2007):

- **44 percent of CH_4** methane,
- **53 percent of N_2O** nitrous oxide,
- **5 percent of CO_2** carbon dioxide.

Plus: Globalized food markets cause many airmiles and employ a major chunk[L] of the large maritime merchant fleets which constantly cross the oceans. The cargo ships burn the dirtiest oil, crude oil, and also release refrigerants (17.5 million metric tons CO_2e annually),[76] and require astronomical amounts of steel for the construction of the ships and containers. And frequent repainting.

Already in the 1970s (!) studies showed that producing the food stuffs for a meat-based diet requires about twice the amount of energy (kcal) than what such a diet yields.[77] (A vegetarian diet breaks even.) This waste of energy was possible over the

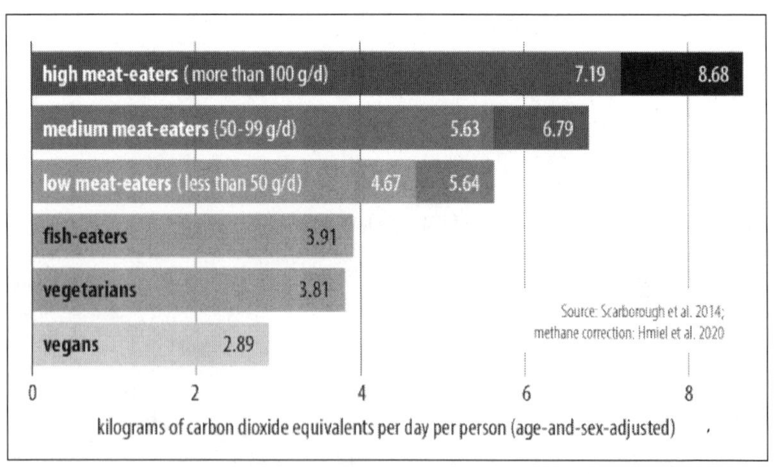

Figure 21. Average greenhouse gas emissions for various diets

The additional higher values for meat reflect the methane correction (from 17% to 25% contribution to global heating).

last half century because of Cheap Oil. But in an era when the biggest global challenge is how to produce (non-fossil) energy, this type of "luxury" is bound to fade out.

In December 2019, over fifty scientists called for governments in all but the poorest countries to renew their pledges to the Paris Climate Agreement to transform agriculture. Because livestock is a significant and fast-growing source of global greenhouse gas emissions, setting a date for Peak Meat would be an urgent but effective measure.[78]

To help reduce the risk of global temperature rising beyond 1.5°C or 2°C, the plea calls on high-income and middle-income countries to incorporate four measures into their revised commitments to meeting the Paris Agreement, from 2020 onwards:

1. Declare a timeframe for peak livestock, i.e. livestock production from each species would not continue to increase from this point forward.
2. Within the livestock sector, identify the largest emissions sources or the largest land occupiers, or both, and set appropriate reduction targets for production. This process would be repeated sequentially, to set reduction targets for the next largest emitter or land occupier.
3. Within a reconfiguration of the agriculture sector, apply a best available food strategy to diversify food production by replacing livestock with foods that simultaneously minimize environmental burdens and maximize public health benefits, mainly pulses (including beans, peas, and lentils), grains, fruits, vegetables, nuts, and seeds.
4. When grazing land is not required or is unsuitable for horticulture or arable production, adopt a natural climate solutions approach where possible, to repurpose land as a carbon sink by restoring native vegetation

cover to its maximum carbon sequestration potential, with additional benefits to biodiversity.

While governments hesitate, millions of people make their own food choices. Numbers of vegetarians are rising, a third of Britons have stopped or reduced eating meat, and veganism is the fastest growing lifestyle movement.[79]

Regenerative agriculture or **carbon farming** comprises an array of techniques that rebuild soil and, in the process, sequester carbon. Increasing the carbon content of the world's soils by only 4 parts per 1,000 (0.4 percent) each year would remove around 3-4Gt of CO_2 from the atmosphere, the equivalent to the fossil fuel emissions of the European Union.[80] Generally, regenerative agriculture methods use cover crops and perennials so that bare soil is never exposed, and let animals graze in ways that resemble animals in nature, i.e. without being overgrazed the grasses have a chance to recover and develop new root mass. Such methods offer ecological benefits far beyond carbon storage: they stop soil erosion, restore a high mineral content in the soils, protect the purity of groundwater and reduce pesticide and fertilizer runoff.

Yields from regenerative methods often exceed conventional yields. And because these methods build soil, crowd out weeds and retain moisture, fertilizer and herbicide inputs can be reduced or eliminated entirely, which saves the farmers money.

No-till methods can sequester up to a ton of (pure) carbon per acre annually (2.5 metric tons/hectare). In the USA alone, that could amount to nearly a quarter of current emissions. According to Rattan Lal of Ohio State

University, such soil regeneration methods could enable desertified and otherwise degraded soils to sequester up to 3Gt of *pure* carbon worldwide every year (equal to 11Gt of carbon dioxide, or nearly one-third of current emissions!).

Other experts see even greater potential: According to studies by the Rodale Institute, organic regenerative techniques practiced on cultivated land could offset over 40 percent of global emissions, while practicing them on pasture land could offset 71 percent. That adds up to land-based CO_2 reduction of over 100 percent of current emissions, not even including reforestation and afforestation, which could offset another 10-15 percent. A bright ray of hope because humanity not only needs to stop ongoing carbon emissions but also to seriously reduce present atmospheric levels.[81]

Abyss of insanity: globalized shipping

The other "overlooked" aspect in calculating anthropogenic carbon output is that *emissions from ships, particularly the merchant fleets, are not even included in the IPCC calculations* or in national compilations of GHG emissions. Governments (as well as the EU) have consistently played down the climate impact of shipping, claiming it to be less than 2 percent of global emissions. Until a leaked UN study showed that in 2007 annual emissions from the world's merchant fleet had already reached nearly 4.5 percent at that point in time, almost twice as much as the aviation industry.[M,82] But even ship engines improve (slightly), and that figure is now fluctuating between 2.6 percent and 3 percent. However, it is still *not* incorporated in the alleged 100 percent of annual global GHG emissions.[83] Only inland water navigation is accounted for by the IPCC.

The type of fuel that high sea ships use, fuel oil, is the dirtiest there is. It is the bottom residue sludge from the refinery bins, heavier than diesel and heating oil, and permitted to contain up to 3.5 percent sulfur. (At least on inland waterways, ships have to use the less heavy marine diesel oil.) A car driving 15,000km a year emits on average 101 grammes of sulfur oxide gases (SO_x) in that time. One of the world's largest ships' engines generates about 5,200 metric tons over a year. Hence a headline went around the world in 2009 saying that "just 15 of the world's biggest ships emit as much pollution as all the world's 760m cars."[84] This is faulty unless you substitute sulfur dioxide for "pollution". (And it's not true anymore because already in 2017 the number of cars in the world passed the one-billion mark.) Apart from CO_2 and SO_x, ship emissions also contain large amounts of nitrogen oxides (NO_x), carbon black, and particulates.[N]

There are 100,000 ships in the world, and more than 90 percent of the world's traded goods travel by sea.[85] The University Corporation for Atmospheric Research (UCAR) has produced a splendid animation video that shows the increasing velocity of the naval trade routes 1800-2013.[86]

The cruise ship sector is underestimated too. It's image of luxury might lead us to believe that taking a cruise is something for the more (or most) affluent people. But the cruise ship sector is huge, and so is its impact on the ecosphere. The commercial boom is remarkable too, passenger numbers grew from 17.8 million in 2009 to **over 29 million** in 2019.[O] All these passengers, plus further millions of crew and staff,[P] are being shipped around mostly on crude oil. Destination harbors are frequented by hundreds of mega ships every year, with 2 to 5 thousand tourists spilling out at a time, hungry for selfies and plastic souvenirs. The fossil fuel engines keep running at the docks because about 40 percent of their output generates electricity for the hotel aspect, the restaurants, and amenity areas on board. In 2017, the cruise ships of the biggest operator alone, Carnival Corp., emitted almost ten times more sulfur oxides along Europe's coasts than all 260+ million passenger cars in Europe combined.[87]

Many of these ships are like small cities, and leave the usual waste in their wake. A large cruise ship, carrying over 7,000 passengers and crew on a one week voyage, is estimated to generate 210,000 gallons of human sewage and 1 million gallons of graywater (water from sinks, baths, showers, laundry and galleys). Modern ships have sewage treatment facilities in their belly, but the discharge is still considerable (also think of microplastics, chemical pollutants, antibiotics, and other pharmaceutica that pass the filters). Plus: tour operators don't always bother with regulations.

Carnival Corp. was on probation after a $40 million settlement (at a Miami court in 2017) for violating environmental laws (its third conviction for the same offence since 1998), but an inspection revealed over 800 further environmental violations during the twelve months following the settlement. Carnival had "illegally released over 500,000 gallons of sewage and over 11,000 gallons of food waste into water near ports and shores around the world."[88] And according to a report by Friends of the Earth,[89] Carnival is not the only cruise line with a criminal record.[Q] Such scandals do not seem to affect the unusually loyal customer base; people believe the adverts about cruising being a "conscious, mindful way of traveling."[90] A new 2019 specialty was to watch the melting of the polar ice caps over a glass of champagne on the cabin balcony.

What homo sapiens can't master, perhaps coronavirus can. When Japanese authorities quarantined Carnival's Diamond Princess off the coast of Yokohama in February 2020, the 712 infections among passengers were the single biggest cluster of COVID-19 cases outside China. Media began to call cruise ships "floating Petri dishes". During spring/summer 2020, all 338 cruise ships currently in existence were docked. Governments had issued "no sail" edicts and the majority of the expected 32m passengers stayed at home.[91] Albeit the engines and systems of these ghost-town ships were kept running to make sure that none seize up.

Post-corona the industry is forced to rebuild public trust, and install social distancing rules: perform pre-departure health screenings, limit onshore day trips, and split the timetable for activities so that only half the number of passengers would take part at any time. It's hard enough for cinemas and restaurants on land to operate at 50 percent capacity, but can cruise ships? There is hope on the horizon for a cleaner, less crowded horizon.[92]

Back to Smoke & Mirrors

Politicians and mainstream media remain reluctant to even mention the climate impact of industrial agribusiness, and of ships; instead the heat is on the comparatively minor polluting sectors: private cars and holiday flights. Very well, they too need to come down in emissions, and everyone needs to start making sacrifices. Only if the voters show a real will for change might there be a small chance that politics could maybe begin to enact regulations to stir this whole big thing around. But there is another, a darker aspect behind it. It's to do with the mirror part in *smoke and mirrors*.

What do you do to prevent, or delay, the population from demanding the industries to cut down on profit in order to empower environmental regulations? You pull out the mirror: "Before *you* complain about the industrial chimney stacks, what about your own exhaust fumes then?!?" It all comes down to the choice of the consumer, right? The 8th of October 2018 was a day of truth, the IPCC conference told the world press about the state of the planet. The core message was: a warming of even 1.5°C (2.7°F) would take humanity into unknown, uncharted and very dangerous territory; humanity has ten or twelve years to radically change carbon behavior. The next day was scheduled for the EU Conference of Interior Ministers. Perfect timing to begin right away on the big steps necessary! But instead, the whole day had only one topic: whether to reduce future car emissions by 40 or by 30 percent (in the end, 35 percent was agreed). No mention of the IPCC's dire warning to transform *the whole* of the economy.

To add another layer, in the months to follow an even fiercer discussion about nitrogen dioxide (NO_2) from diesel engines was pushed. Of course, the car industry shall be brought to justice about the emissions scandal (dieselgate)! But the technology to fix this problem exists, and the budget is there too. *In terms of climate, there is no NO_2 crisis.*[R] To curb air pollution, some cities enacted movement restrictions for diesel cars but ignored that the annual emissions (nitrogen oxides, sulfur oxide) from cruise ships in their harbors are orders of magnitude higher than those from road traffic.[93]

So much for our example on distraction. The point is: nitrogen oxide emissions from diesel cars have little to do with global heating, but headlining them serves to burn out the public energy for conservation issues. Nobody has got the stamina left to even start challenging Big Oil and Big Food. This was a masterpiece of smoke & mirrors. And it ran well, right until the COVID-19 measures presented themselves as the next opportunity to create an efficient smoke screen. The recipe being: whenever people begin to care for the planet, force them back to separation and fearful navel-gazing.

While corporate business is generally very protective of itself, there is a surprising degree of accordance with the notion that the car sector is to blame for everything. This raises the question whether the Big Oil lobby, in its last phase of resistance against the global shift to non-fossil energy, is possibly playing the private motoring branch as a *pawn sacrifice*. A move no one would anticipate. Decades of public discussions about the evils of private motoring and ever-new generations of still-useless hybrid or electric cars just distract so well from the vast changes that industrial society as a whole and particularly oil-based agribusiness need to make.

Electric cars

Electric cars have finally begun to improve, but their overall ecological footprint is still huge, and battery technology is neither green nor fair. Hybrids combine the worst of both worlds, due to the extra weight of a second engine, no existing model (in 2020) has an ecological footprint that outperforms a conventional gasoline or diesel car[94] (with the only exception of short distances in cities—for which a bicycle might be better anyway).

The true ecological and work-ethical price for electric cars is still appalling. Manganese, cobalt, graphite and lithium needed for the batteries are mined in Africa and Chile where the industrial process uses vast amounts of water. The farming soils in these regions (like the Atacama Desert in Chile) go dry. Cobalt in the DR Congo is mined under slavery-like conditions, and with no breathing protection; the miners' villages have only contaminated water and no trash collection. Copper production creates huge amounts of poisoned water. The car industry needs hundreds of thousands of tons of copper for wiring and motor parts: this doesn't change if a car runs on gasoline or electricity. Apart from that, much of the electricity for electric cars currently still comes from unsustainable sources such as coal or wood pellets facilities.[95] And last not least, tire abrasion remains a large problem with all vehicles.

Facing the climate emergency, electrification sadly is needed, but a big part of any solution is a much reduced vehicle fleet: away from the private car. As two-thirds of car trips are under 5 miles, encouraging walking and bicycle use (like very successfully done in Copenhagen) is a starting point. Other measures are much improved public transport, affordable taxi and car sharing schemes, and special fleets for senior citizens and others with mobility issues. City planners in Portland OR, in Melbourne and in Paris, were the first to restructure their cities for a better future. "Green transport policies," as the deep green ecologist Sandy Irvine points out, "also improve health and fitness, raise the quality of life, build community cohesion, reduce inequalities, and unlock land now lost to road and parking spaces."[96]

A promising way to store electricity without the devastating ecological impacts of lithium production is hydrogen technology. Sadly, UK governmental funds for battery research (lithium, etc.) are ten times higher than for hydrogen technology (£25m over three years).[97]

Methane

What we also really have to look at, apart from CO_2, is methane (CH_4). Since 1990, the increase in atmospheric GHG levels has made the heating effect of the atmosphere 43 percent stronger. Until early 2020 it was believed that four-fifths of that is caused by CO_2, and 17 percent by methane. Methane is dominantly produced by livestock, and also by oil and gas extraction (many fossil fuel plants routinely vent methane gas into the atmosphere,[98] and fracking has massively increased methane in the atmosphere since the millennium), by tropical wetlands, and, to a much smaller extent, by landfills. The third anthropogenic GHG is nitrous oxide (N_2O), commonly known as laughing gas and not to be confused with nitrogen dioxide (NO_2) from cars, which stems from heavy use of artificial fertilizers and from forest burning. Its atmospheric levels are comparatively low but as a GHG it is around 300 times more potent than CO_2. Its concentration in the atmosphere has been calculated as now being 23 percent higher than in the mid-eighteenth century, before the Industrial Revolution.[99]

Methane's global heating potential is 86 times higher than that of CO_2 over a 20-year period and about 34 times higher over 100 years. Its atmospheric concentration is now more than double pre-industrial levels. New discoveries published in *Nature* in February 2020 revealed that the share of *naturally*

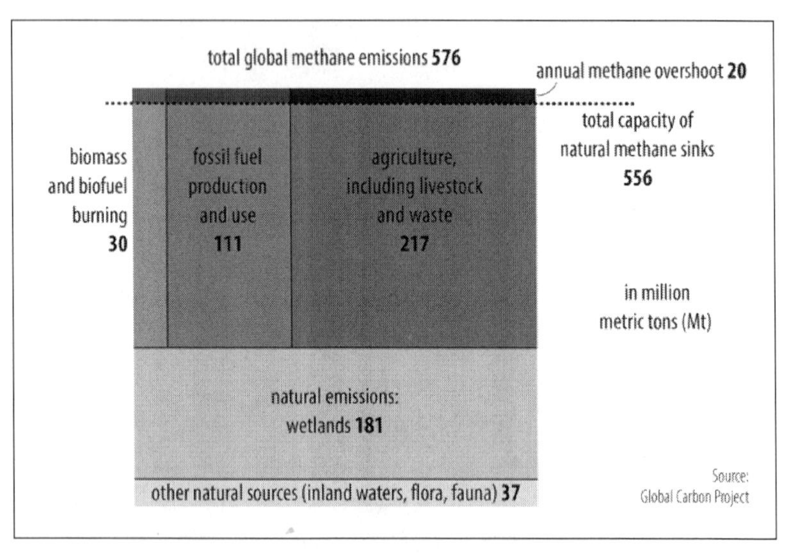

Figure 22. Global methane budget 2008–2017

The top slice shows the annual overshoot of methane in the climate system, representing the degree by which human activities overburden the planet's capacity to remove methane from the atmosphere. And with the ongoing degeneration of the natural world this capacity is diminishing.

released fossil methane had been overestimated by "an order of magnitude".[100] This means that human activities are 25-40 percent more responsible for fossil methane in the atmosphere than previously thought. An earlier study[101] uncovered that methane emissions from US oil and gas plants were 60 percent higher than reported to the EPA (Environmental Protection Agency). *Methane is now believed to be responsible for about a quarter of global heating.* This puts much more pressure on the excavation processes of the fossil fuel industry, and on industrial agriculture, particularly the livestock sector. All the more so as The Global Methane Budget report states that the world reached a new methane emissions peak: anthropogenic methane release in 2017 was almost 600m metric tons, which is 50m metric tons more than the annual average a decade before.[102]

Tipping points and Hothouse Earth

The IPCC does not incorporate the chances of tipping points and cascading effects. Quite understandably because a) they make everything incalculable, and b) *nobody has the slightest clue when any of the tipping points may be reached*—or, for that matter, if we already have reached a point of no return and just don't know it. The danger is that by not accounting for them all warnings will remain over-optimistic and naïve.

Reaching a tipping point means overstepping a boundary (previously unknown to us until then) in the balance of Earth's life support systems beyond which cascading effects are set into motion like a domino chain. This would lead to runaway heating that could not be influenced by humans anymore. The end result would be the so-called "Hothouse Earth".

The Harvard study of that name[103] states that "precisely where a potential planetary threshold might be is uncertain. We suggest 2°C because of the risk that a 2°C warming could activate important tipping elements," namely "intrinsic biogeophysical feedbacks in the Earth System" which could lead to initiating the *irreversible* Hothouse Earth pathway as early as 2030. Hothouse Earth is often understood as being 5°C or 6°C above pre-industrial levels. But even 3°C most probably leads to the collapse of most regional weather systems such as the monsoons of India and Africa, and the Gulf Stream > leading to epic losses of harvest > collapse of the global food system > loss of civilization as we know it. And even 1.5°C might be enough to cause a serious methane "burp" from beneath the Arctic Sea (see "Conclusion", below).

Evidence suggests that global tipping points could be triggered between 1°C and 2°C. This risk is "an existential threat to civilisation," says Prof. Tim Lenton at the University of Exeter, meaning "we are in a state of planetary emergency."[104]

The most important tipping elements are:

- the melting of the summer ice in the Arctic Sea,
- the melting of the Greenland ice sheet,
- methane release from thawing permafrost soils in the Arctic,
- methane release from beneath the Arctic ocean bed,
- the melting of the ice shield in western Antarctica,
- the warming of the oceans and the weakening of their carbon absorption,
- collapse of the tropical rainforest,
- the weakening of thermal circulation in the Atlantic and thus the Gulf Stream,
- destruction of coral reefs.

Other tipping elements of global importance are:

- changes of the El Niño-Southern Oscillation (ENSO),
- collapse of the Indian summer monsoon, and of the West African monsoon system,

- retreat of the boreal forests (Canada, Scandinavia, Russia),
- disappearance of the Tibetan glaciers,
- desertification of the North American southwest.

First tipping point example: the oceans absorb about one-third of man-made CO_2 emissions. This creates carbonic acid, and leads to ocean acidification, and current levels of acidity are underway to threaten and diminish marine organisms who still work hard to counteract the carbonification of Gaia's domains (see "The carbon cycle", Chapter 3).[105] With each carbon-sequestering plant population disappearing, the sea loses its potential to absorb carbon. The positive feedback cycle is: when marine life dies back > the sea acidifies even quicker > it takes less CO_2 from the air > the planet warms faster > warmer water cannot hold so much carbon as colder water, hence the saturation increases > even more marine life dies. This might seem to be a steep linear decline towards hitting rock bottom, but we cannot expect it to be even (and hence mathematically predictable) because somewhere along the way there is a tipping point waiting that will suddenly take the collapse to rocket-speed. This would be a tipping point affecting the system itself.

But there is another level: when the sea is completely saturated with carbonic acid, not only will it *stop* absorbing CO_2, a warmer sea will begin to *release* what it had accumulated in the past. This would be a tipping point of global ramifications––meaning that there would be no more brake in human reach anymore; even if humanity had already stopped emitting carbon, the planet would spiral into a hot state.

Second example: the Amazon rainforest still absorbs about a fifth of the world's total carbon emissions, but this crucial ability is decreasing fast. Global heating causes droughts even in the Amazon, and interestingly they are worse in regions with a lot of clearfelling. Lack of water inhibits plant growth > less growth means less carbon is being built into plant structures > less carbon absorption by the forest leads to more global heating > warming causes more severe droughts. A major drought affected a third of the Amazon in 2005, and an even worse one affected half the Amazon in 2010.[106]

When the Amazon ecosystem becomes too fragmented, there is a tipping point at which this gigantic global water pump will simply stop working. Within half a decade the continent would turn barren, and the further loss of biodiversity would be beyond our comprehension. So would be the scale of human suffering, and the size of the migration caravans. And that's not all, the collapse of the Amazon would have intercontinental implications: the Gulf Stream might become seriously impaired, exposing North America to more severe droughts and Europe to unprecedented coldness. The dead rainforest would begin to release the carbon stored in its wood, possibly more than 50 billion metric tons over 30 to 50 years.[107] The annihilation of this gigantic carbon sink (including the release of methane and nitrous oxide by the burning of the vegetation) and the subsequent chain reactions would be a deathblow to global climate and probably lead to Hothouse Earth. 17 percent of the Amazon rainforest has been lost since 1970. The tipping point has been calculated to lie in the range of 20-40 percent, so we are coming close.[108] At the current rate of deforestation, the tipping point for the collapse of the Amazon rainforest is estimated to be reached between 2040 and 2070.[109]

By the way, faster tree growth due to higher atmospheric CO_2 levels is not occurring, contrary to some expectations. This is mainly due to effects of global heating, predominantly heat and lack of water.[110]

Third example: the Gulf Stream. The warm Gulf Stream is part of the Atlantic meridional overturning circulation (AMOC), mighty currents that bring warm Atlantic water north towards the pole cool in the Arctic region, sink and return southwards. This system has a major impact on the climate of the northern hemisphere. With the rapid warming of the polar region, the AMOC are beginning to slow, and with them the Gulf Stream. The warming of the Arctic could eventually lead to the full failure of the Gulf Current, warns oceanographer Dr. Ira Leifer: "The resultant deep freeze that would hit Europe would destroy European agriculture and likely lead to a massive war for survival."[111] Oceanographer Peter Spooner, at University College London, is concerned too: "The extent of the changes we have discovered […] points to significant changes in the future."[112]

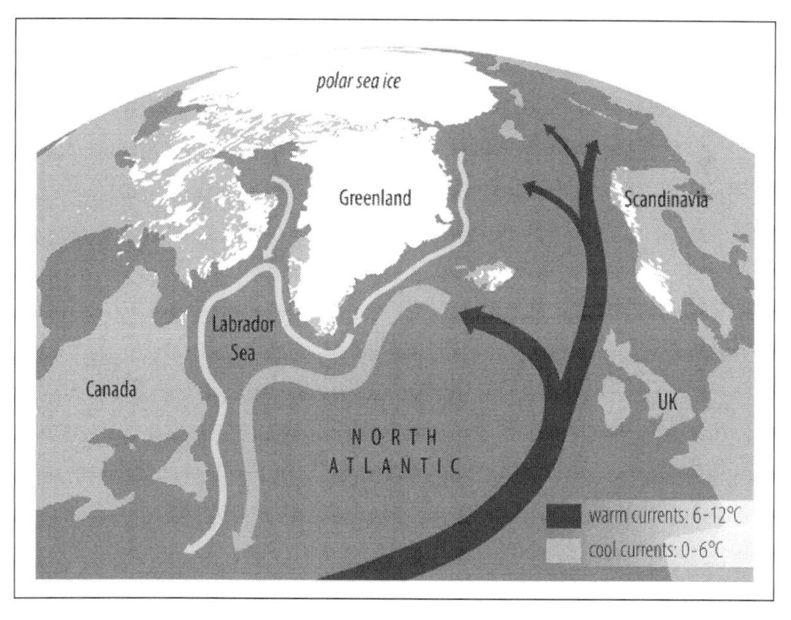

Figure 23. The Gulf stream

The jet stream[113]

Jet streams are fast flowing, meandering air currents in the atmospheres of some planets, e.g. Jupiter and Earth. On Earth, each hemisphere has a *polar* and a somewhat weaker *tropical* jet stream. The polar jet streams develop in airy

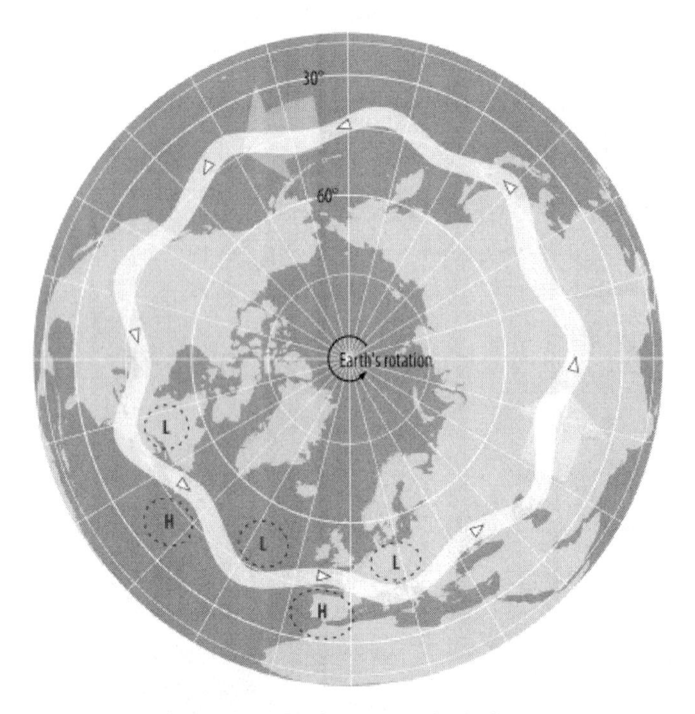

Figure 24. The jet stream in balance

The northern polar jet stream is a band of westerly winds (flowing west to east) between 30 and 60 degrees north. In the summer, when subtropical and polar air temperatures are less different than in winter, this band can develop meandering curves. The jet stream is the weather maker: low pressure areas develop north of the stream, high pressure areas on the south side. These weather systems slowly drift eastward, along with the meandering shape of the jet stream.

Figure 25. The jet stream changes

Along with global climate disruption, the jet stream becomes much wider in amplitude. The whole system loses strength and the entire eastward drift slows down, or even comes to a halt and locks regional weather conditions in place. On 22 July 2018 this caused a day of devastation around the world: (1) wildfires in California, (2) a heatwave across the Southwest, (3) floods in the northeastern USA, (4) a heatwave in Scandinavia, (5) drought across Central Europe, (6) wildfires in Greece, and (7) a heatwave in Japan.

regions where warm air from the subtropics moves towards a pole and meets cold air from the polar region. The greater the temperature difference when subtropical and polar air meet, the more dynamic the jet stream becomes.

The northern polar jet stream on Earth occurs at a height of 27,000-39,000ft (8-12km) above sea level, and is usually a few hundred miles or kilometers wide. Within the stream, wind speeds range from 57mph (92km/h) to as much as 250mph (400km/h). They are always westerly winds (flowing west to east).

Usually, the path of the jet stream has a meandering shape (Rossby waves), and these meanders themselves drift eastward, at lower speeds than those of the actual winds within the flow. The polar jet stream is the weather maker of the northern hemisphere: low pressure areas develop north of the stream, and high pressure areas on the south side. And these weather systems drift along as the whole system propagates eastward. This is part of the benevolent climate that humanity has enjoyed for the last ten millennia.

Because the temperature differences between polar and subtropical regions are smaller in summer than in winter, the current is weaker and the Rossby waves tend to develop extended north-south curves. This is because cold air masses in the north and warm ones on the south side usually act like a wave duct funneling the Rossby waves around the globe.

But with temperature differences decreasing due to global heating, the borders of this atmospheric "riverbed" erode and the Rossby waves become much wider in amplitude. With this dispersal of their strength, the entire eastward drift slows down, and sometimes even comes to a complete halt. At that point, regional weather conditions get locked in place: torrential rains don't move on as they used to, but stay for days or weeks, so do heatwaves at other places. And, interestingly, these extreme weather types occur at the same time. For example:

- July 2010: record heat wave in Russia leading to drought and wildfires, while century floods drown large areas in Pakistan;
- July 2018: heat wave in the southwest of the USA and simultaneous flooding in the northeast (especially in Pennsylvania and Maryland);
- 2007 and 2012: severe flooding in the UK as a result of the polar jet being locked for the summer;
- June/July 2020: drought in Europe, heatwave and wildfires across Siberia, while torrential rains along the Yangtze River lead to the 2020 China floods.

Climate scientists expect that the jet stream, like the Gulf Stream, will continue to weaken as a result of global heating.[114] As a consequence, weather extremes are becoming much more frequent. This includes extreme winter weather, because with a weaker jet stream, the Polar vortex has a higher probability to leak out and bring extremely cold weather to the middle latitude regions, which at other periods will have extreme heatwaves coming from the south.

All factors indicate that extreme weather phenomena are becoming increasingly severe. A warmer atmosphere stores more moisture, resulting in heavier rainfall and flooding. And a warmer atmosphere means more frequent, longer and more intense heat waves.

The final example: the Arctic methane. A giant has begun to stir beneath the thawing permafrost soils in the Arctic circle. For a few years now unusual warm weather and heatwaves in Siberia have time and again resulted in local but violent explosions of methane gas bursting out of its subterranean deposits. Overall, methane levels over land are slowly rising. Methane (CH_4) is of huge concern because it is about 84 times more potent a greenhouse gas than CO_2. The twin to the mainland methane deposits are the methane hydrate reservoirs under the offshore seabeds of the East Siberian Arctic Shelf (ESAS). A 2017 study[115] led by Natalia Shakhova from the International Arctic Research Center at the University Alaska Fairbanks shows that the geological layers that cover them are seismically and tectonically unstable; vents provide gas migration pathways already. The water in those areas is only 50 meters deep, and the ice cover with its high albedo is disappearing > water absorbs more sunlight and warms quicker than ever before. On top of that, test drilling around the Arctic coasts is an insane provocation of this fragile balance, and is causing numerous methane gas blowouts, both on-land and offshore. At the conference of the European Geosciences Union in 2012, Prof. Shakhova explained: "The current CH_4 level in the atmosphere is about 5Gt [billion metric tons], while the volume of CH_4 within the East Siberian Arctic Shelf (ESAS) is estimated to be between hundreds and thousands of Gt. And of course, only 1 percent of that amount is required to double the atmospheric burden of methane. To destabilize 1 percent of this carbon pool does not require much effort, [...] There is a potential risk for this to happen." Shakhova is worried about the Arctic Shelf: "We do not like what we see."[116] The East Siberian Arctic Shelf might turn out to be the Achilles' heel of the global climate system.

Conclusion

What is being done?

• The UN Climate Conference (COP24, December 2018) led to an agreement (that's the good news) on unified ways of measuring the damage and gathering the data. But no path of action! All that the UN climate council can do is to *recommend* reducing destruction. Nothing is or will be legally binding.

And never forget that while governments faithfully attend the stage of the regular climate conferences (voters like to see the good will), at home they still pay fortunes in oil and fossil fuel subsidies: the governments of the G20 are estimated to be spending $88 billion every year subsidizing exploration for fossil fuels.[S] On top of that: in June 2019, as the worst wildfire season ever began to ravage the Amazon rainforest ("Land Use Change" for the livestock market), European leaders signed the Mercosur deal with their South American counterparts, ensuring even more soy and beef imports to Europe, in return for cars and agro-chemicals for South America—a spit in the face of the climate movement. It really is high time to lose patience!

• Various political players such as the UN, the EU, the Democrats in the USA, and Labour in the UK had begun to discuss a Green New Deal (see box) when the American politician Alexandria Ocasio-Cortez, a fervent supporter of this concept, was voted into the US Congress in 2018. In April 2020, the European Parliament called to include the European Green Deal in the recovery program from the COVID-19 crisis: "The European Green Deal and the digital transformation should be at its core in order to kick-start the economy, MEPs stress."[117]

A serious, radical Green New Deal would implement a consequent declining cap on fossil fuels as a necessary step to constrain global heating and stabilize the ecosphere, but doing so will face massive resistance from Big Oil and Big Coal. They will use their astronomic funding and lobby influence on governments to avoid any serious attempt to regulate fossil fuels. "Therefore, I expect," writes Stan Cox in his book *The Green New Deal and Beyond: Ending the Climate Emergency While We Still Can*, "that a Congress determined to impose a cap and prevent climate catastrophe would have no choice but to nationalize the fossil fuel industries, and the states would need to convert private gas, water, and electric (including renewable electric) utilities into locally controlled public utilities."[118] It is becoming obvious why the fossil fuel industry is so adamantly resistant to systemic change.

• In 2015 at the Paris climate summit, France launched the 4p1000 initiative—to promote research and actions globally to increase soil carbon stocks by 4 parts per 1,000 per year.[119]

The Green New Deal aims to address ecological breakdown and economic inequality *together*. It is very compatible with the UN Sustainable Development Goals due to its focus on social issues (income, affordable health services, education, equality, inclusion, etc.) as well as ecological challenges (climate breakdown, mass extinction, pollution, overshoot). Taxing the rich (corporates) would create sufficient funds to pay Universal Basic Income (UBI)[120] to everyone (which is the only fair social path anyway, regarding the mass digitalization of work expected for the coming years), and also train and fund people to refit their homes and make them more energy-efficient. At its best, a Green New Deal would address all aspects of the ecological crisis in an adequate and bold way. However, it may get watered down or become distorted, e.g. by merging with the neoliberal belief in high tech fixes, smart farming, and the "powers of the markets".[121] At its core, a true Green New Deal could not support the belief in endless economic growth, but would rather promote Degrowth (see box at the end of Chapter 13) and decentralization.

The UN Sustainable Development Goals (SDGs)[122] have been hailed the most ambitious plan of humanity and *seem* to be the best shot we've got to save our species and what's left of the bereaved ecosphere. The 17 SDGs and their 169 targets aim to globally "end poverty and hunger", to "foster peaceful, just and inclusive societies which are free from fear and violence", and "to protect the planet from degradation, including through sustainable consumption and production, sustainably managing its natural resources and taking urgent action on climate change, so that it can support the needs of the present and future generations."[123]

However, the list of these 17 Goals reveals at first glance that the entire plan is rather anthropocentric, i.e. more about people than the planet. It begins with the unclarity of the terms themselves: is "sustainable" really meant in a truly ecological way or rather economically? And do you hear the alarm bells when the word "development" is mentioned? Does it not usually refer to "progress" and "growth"? Indeed, Goal 8 is dedicated to "decent work and economic growth." The SDGs have been criticized for their inherent and unsolvable contradictions. A more decisive stand for the Earth is lacking.

Goals 1 to 11 are openly anthropocentric: poverty, hunger, health, education, gender equality, clean water, clean energy, etc. Of course these Goals are all sorely needed, fair and just, and they will also have positive repercussions on wild nature. But only a minority of three Goals focuses on the ecosphere: Goals 13 Climate Action, 14 Life below Water, and 15 Life on Land. All targets under Climate Action solely deal with the *effects* of climate disruption, not a word about *preventing* further

GHG emissions, not a word about climate *care*. They only recommend to make nations (particularly in the Global South) and islands more resilient for the bad tidings underway. Goals 14 and 15 talk about "sustainable" fishery, or forestry, respectively, which would be a step forward, but the whole agenda is written and conceived in a spirit of *utilizing ecosystem services*, thereby revealing the old anthropocentric attitude that the whole community of life on Earth is only here to serve *us*.

Summary: The Agenda 2030 for Sustainable Development is an important tool to get decision makers moving towards the necessary systemic change. But it is still making concessions, no, it *is* one single large concession to our outdated model of an extractive economy. It doesn't need improvement but a serious overhaul. We have to hope that the discussion will open to a more honest and radical commitment to all life on Earth, and to respect nature's laws and to live in harmony with them.

"Development"

In his textbook on *The History of Development*, Gilbert Rist exposes "development" as an element of modern Western myth. We are better, we know better, and who else could tell the others what to do. It might be (increasingly) disguised as human aid or nature conservation, but in the end of the day it is about arrogance, investment return and economic growth. It is an aspect of a society self-perpetuating itself and its values, stretching out to expand its territory.

The concept of "development" has become part of

the Western belief system. And belief systems are rarely reflected upon. Who is to judge which country needs to be "developed"? Rist's warning therefore is "that we do not yield to ready-made appraisals deriving from the presuppositions of conventional thinking, which would make us take it for granted that 'development' exists, that it has a positive value, that it is desirable or even necessary."[124] Indeed there are many societies in the world, in Africa, South America or in rural Ladakh, for example, where people have much less but are infinitely more happy than people in industrialized countries where they are being reduced to mere consumers and deeply exhausted by the never-ending competitive chase for the golden goose.

The idea of "development" entered the political arena in 1949,[T] in the course of restructuring a world in disarray after World War II. As racism couldn't justify exploitation any longer (the Nazi concentration camps had left a stark warning where racism can lead), Western industrialized countries soon declared themselves to be "developed", and by doing so set the moral, social, and economic standards what "developed" and "underdeveloped" have to mean for the rest of the world. Since then, every success in developing "poor" countries—which are poor in the first place because Western intrusion, invasion, and domination plunged them into the abyss some two centuries ago—is measured against the material prosperity of the rich countries. But the wealth of the latter is still, and more than ever, based on the ruthless extraction of resources from the former, and the militaristic and economic reality of this always wins over the humanitarian hopes to bring true and durable well-being to all humans on Earth. The triumph of neoliberalism and globalization during the last four decades finally cemented "a system which maintains and reinforces exclusion while claiming to eliminate it" (Rist[125]). Only now do growing inequalities and the global ecological meltdown—as the ability of the ecosphere to absorb the detrimental effects of human activities is reaching its limits—force a rethink of the global system, and "the final chapter takes up the debate on 'degrowth' and 'downscaling'."[126]

The ecological overtones of "development" began with the Brundtland Report (commissioned in 1983, published in 1988)[127] which remarked that the deterioration of the ecosphere could not continue much longer, and therefore claimed that "what is needed now is a new era of economic growth—growth that is forceful and at the same time socially and environmentally sustainable."[128] From this grew the United Nations Conference on Environment and Development (UNCED) and its main projects: the Rio Declaration (the "Earth Charter"), The Convention on Climate Change, The Convention on Biological Diversity, and Agenda 21, more recently morphed into Agenda 2030 and its Millennium Goals. But all along, the key to the music remains the same: it is anthropocentric, and "sustainable development" rarely means care for the ecosphere, but growth everlasting, while averting any radical change. Rist once more: "There should be no illusion about what is going on. The thing that is meant to be sustained really is 'development', not the tolerance capacity of the ecosystem or of human societies."[129]

Lacking clear official definition, the terms "sustainability" and "development" mean to anyone whatever they want them to mean. This deliberate ambiguity is the reason for the widespread success of these terms.

Since the millennium, the "end of the century" (2100) is mentioned a lot. Its soothing effect—It's so far away!—calms everybody (which is why politicians and the industry still like to use it). That might have been acceptable for vague predictions in the 1990s but not anymore. Whenever you still hear this number in an ecological context (unless it's about population estimates) be prepared to read 2050 instead.

At the current rate of beating around the bush we are nowhere near on track for 2°C, let alone 1.5°C of global heating, but rather for 3°C. We already reached 1°C in 2018. And it wasn't an *equal* rise since 1950 either: the last third of warming (from 0.7 upwards) only happened in the last four years.[130] 2°C warming will have major ramifications for societies (human, animal and plant alike), as the 2018 IPCC report says. Heating the planet beyond this would create a "totally different world," says Michael Oppenheimer, a climate scientist at Princeton University. "It would be indescribable, it would turn the world upside down in terms of its climate. There would be nothing like it in the history of civilization." Without beginning to make major changes by 2020, a warming of around 3°C is more likely.[131] Even 4°C is realistic, according to scientists at the Swiss Federal Institute of Technology in Zurich.[132] Bottomline: without drastic action and serious regulation we are headed for Hothouse Earth.

There is actually nobody to guarantee that even 1.2°C could not be the trigger for the Arctic Sea methane giant to break out. The sea floor is thawing and loosening as we speak. A cascade of global tipping points triggered by only 1 percent of the Arctic methane could rocket the planet into the Hothouse state, possibly even wiping out all life on Earth bar certain bacteria within a decade. If we don't dramatically reduce CO_2 emissions, pesticides, and habitat and biodiversity loss during the very early 2020s, the world as we know it will not be the same by 2030. And how will we live after the collapse of the Amazon? What will Europe be like without the Gulf Stream or agriculture? In which parts of the world might humans fall victim to an MME (see Chapter 9). And why should we even want to go anywhere near finding such answers? Should we not have long ago employed the *precautionary principle* and pulled the brakes?

But we are still rushing forward, always reaching for the bright dangling carrot called "progress and growth". Why do we go this far, blindfold on a rope? We have all the technological and logistical means to steer this runaway train away from the cliff, put the brakes on and choose the other track at the switch. There is such a beautiful, dignified future on a healthy planet waiting for humanity and all more-than-human life. It seems that the core problem is not the inflexibility of the industrial behemoth (because we *could* change it) nor a supposed "evilness" of those very few who are in control of it, but the underlying values and paradigms that *all of us* uphold. The problem is in the head. In everyone's.

Footnotes

A. A large steep slope in the Barry Arm fjord in Alaska has begun to move. Anytime in the next twenty years, its total volume of about 650 million cubic yards (500 million cubic meters) will crush into the fjord and create a tsunami. A similar landslide in Lituya Bay was triggered by an earthquake in 1958, a volume of rock 16 times smaller than that at Barry Arm caused a tsunami wave a few hundred yards high. Barry Arm is now monitored 24/7 throughout the summers by airborne and satellite motion data analysis.[133]

B. Compared to a total of 6,800 square miles (17,600 square kilometers) in January-June 2019.

C. Other ecological implications listed in the 1986 Shell Oil study:

- the disappearance of the coral reefs,
- an existential threat to low-lying countries like Bangladesh,
- an increase in "runoff, destructive floods, and inundation of low-lying farmland",
- "drastic changes to the way people live and work", due to changes in air temperature.

This report also said that the rise of average temperatures would become obvious around the turn of the century, and *that once global heating does become measurable it might be already too late to realize effective countermeasures or even just stabilize the situation.*

D. In this case, 2°C means above the early 1980s levels (and even more compared to pre-industrial levels).

Such early multi-year studies had been triggered by a note from James F. Black, a top technical expert in Exxon's Research & Engineering division, to his company leaders. The memo from July 1977 states: "There is general scientific agreement that the most likely manner in which mankind is influencing the global climate is through carbon-dioxide release from the burning of fossil fuels."[134] And in 1978, Black warned other Exxon scientists as well as managers that a doubling of the CO_2 concentration in the atmosphere would increase average global temperatures by 2 to 3 degrees Celsius (4 to 5 degrees Fahrenheit), and as much as 10 degrees Celsius (18 degrees Fahrenheit) at the poles. He spoke about the chances of rainfall getting heavier in some regions, while others might turn to desert. "Some countries would benefit but others would have their agricultural output reduced or destroyed."[135]

A warning about the global effects of fossil fuel emissions

appeared on American TV as early as 1958.[136]

E. Anti-climate-change lobbying: Since 1998, Exxon alone has given over $31 million to organizations and individuals blocking solutions to climate change and polarizing and misinforming the public.[137] After publicly pledging to stop funding these climate denial groups in 2007, Exxon nevertheless paid more than $2.3 million to organizations and individuals who denied the expert climate consensus and acted to obstruct climate policies.[138] And it continued: an investigation by InfluenceMap showed that US Big Oil "spent almost $115m per year on obstructive climate influencing activities, the bulk by the American Petroleum Institute ($65m), ExxonMobil ($27m) and Shell ($22m)."[139]

   Funding scientists to dispute the global warming consensus has gained momentum since the turn of the millennium: Exxon gave over $1 million to a famous climate change denier,[140] who was also one of the fore-front climate deniers receiving large sums from coal companies like Peabody Energy or Alpha Natural Resources.[141] Analysis of Peabody Energy court documents show company-backed trade groups, lobbyists and think tanks dubbed "heart and soul of climate denial", as well as some well-known climate-denial extremists.[142] Alpha Natural Resources also paid attorney Chris Horner who made himself a name for witch-hunting climate scientists.[143]

F. Climate scientists demanding an investigation were joined in 2015 by a petition with over 350,000 signatures, including Democratic politicians such as Hillary Clinton and Bernie Sanders.[144] The New York State Attorney General was the first to subpoena Exxon in November 2015 for any documents related to its climate science research and communications to shareholders and the public.[145] Since then, Texas, Massachusetts and the US Virgin Islands have followed. Faced with the devastating consequences brought by rising sea levels, eight cities and counties in California, along with New York City and municipalities in Colorado and Washington state, have filed civil lawsuits against several oil and gas companies. Furthermore, since shareholders were also misled for decades,[146] shareholder associations are starting to demand justification, also for the (above-mentioned) annual $115m of shareholder funds for disinformation campaigns.[147] And because of the fundamental betrayal of the younger generations, so far nine children's lawsuits supported by Our Children's Trust have been filed in state courts from Alaska to Florida.[148]

On the other side of the Atlantic, Dutch environmentalists have begun to sue Royal Dutch Shell.[149] This is only the beginning of litigation for this treason against humanity and all of planet Earth. The activist Bill McKibben, founder of 350.org, says, just think of how the world would be different if Big Oil had told the truth, the whole truth, and nothing but the truth on climate change.[150]

G. The elite scientists mentioned were the "Jasons", and their report *The Long-Term Impact of Atmospheric Carbon Dioxide on Climate* was sent "to dozens of scientists in the United States and abroad; to industry groups like the National Coal Association and the Electric Power Research Institute; and within the government, to the National Academy of Sciences, the Commerce Department, the E.P.A., NASA, the Pentagon, the N.S.A., every branch of the military, the National Security Council and the White House."[151]

H. Especially Bush's Chief of Staff, John Sununu, who foiled US action on climate change for good in 1989.[152] Another example: a memo from communications strategist Frank Luntz leaked in 2002 advised Republicans: "Should the public come to believe that the scientific issues are settled, their views about global warming will change accordingly. Therefore, you need to continue to make the lack of scientific certainty a primary issue in the debate."[153]

I. To create distracting smoke requires a good eye for timing. Two examples. Do you remember the media furor when the judge Brett Kavanaugh, whom Trump had positioned for the Supreme Court, was challenged in an elevator by two of his sexual assault victims? The very day (September 28, 2018) that all media eyes were on the elevator instant[154] was the day when a major environmental impact statement was released by the US Dept. of Transportation to the White House. It summarizes that at the present trajectory of GHG emissions, the world will be even 4°C (7°F) hotter in 2100. Finally, Trump had to admit to climate breakdown, but the world looked to the elevator.

The results of the Environmental Impact Statement justified Barack Obama's draft to bring in higher efficiency standards for cars and industry in 2020, with steeply decreasing emissions until 2025. But the Trump administration decided to interpret the results the other way: the planet's going to boil at 4 degrees warmer, so we might as well not bother. "The analysis assumes the planet's fate is already sealed," noted *The Washington Post*.[155] And the Trump administration continued to support aging coal plants, allow oil and gas operations to release more methane into the atmosphere, and scrap Obama's vehicle regulations. The smoke screen tactic was pointed out by several scientists who had worked on the report, according to *The New York Times* they "noted that the timing of its release, [...] the day after Thanksgiving, appeared designed to minimize its public impact."[156]

J. The "cool" phase 1998-2012 confused many people, and put a lot of wind into the sails of climate deniers. Soon after, scientists found various parameters in the climate system that may account for this cooling phase, including a heat transfer into deeper Pacific waters via convection due to the influence of the Pacific Decadal Oscillation (PDO); effects of

La Niña; and movements of the Pacific trade winds which supported this heat exchange in the ocean; also small volcanic eruptions since 1999; and reduced solar activity.[157]

K. The diagram in Figure 17 shows the frequency of Ice Ages. During our present geological period, the Ice Ages have occurred fairly regularly, about every 100,000 years: a rhythm believed to be triggered by the eccentricity of Earth's orbit (see box "Irregularities", Chapter 2).

L. Numbers are hard to find, but here is one example: US grain exports to Asia in the first ten months of 2013 comprised of more than 470,000 twenty-foot containers (doubled since 2006), according the US Department of Agriculture (USDA).[158]

M. Annual emissions from the world's merchant fleet in 2007 were 1.12Gt of CO_2, aviation industry 650mt.[159] But aviation increased and by 2018 emitted 918Mt of CO_2, 2.4 percent of all CO_2 emissions (747Mt for passenger transport and 171Mt for freight operations).[160] Cargo trade is also increasing and emissions from international shipping were expected (before COVID-19) to grow between 50 percent and 250 percent by 2050.[161]

N. To be fair, shipping engines have improved, and in 2007 they used only half as much fuel per tonne of freight than twenty years prior. And transport by cargo ship generates much less GHG emissions per weight and distance than transport by trucks or airplanes. But on the other hand, shipping trade has increased dramatically.

Since 70 percent of all ship emissions are within 400km of land, and 85 percent of all ship pollution is in the northern hemisphere, from 1 January 2015, ships transiting coastal waters have to burn fuel oil with a sulfur content of no more that 0.1 percent (MARPOL agreement). This was decreed to somewhat protect the health of people living on the Continents.[162] Still, it is believed that ship emissions worldwide account for 400,000 early deaths and 14 million asthma cases among children.[163] The Danish health service alone estimates that it spends almost £5bn a year, treating health problems due to shipping emissions.[164]

PS: One simple way of reducing emissions: from 2008 to 2012, due to the financial crisis, ships traveled with only half steam—which saved the world half the emissions (BBC News).

O. Most of the cruise passengers in 2019 were from the USA (45 percent), followed by China, Germany and the UK (9 percent, 8.2 percent, and 7 percent, respectively), followed by Australia, Canada, Italy, Spain, France, Brazil, and others.[165]

P. Since cruise ships are basically floating hotels with shopping malls, a ship with 2500 guests may have almost one thousand staff traveling along (the most luxurious ships even have more crew and staff than passengers).[166]

Q. The others are Seabourn, P&O, Costa, Princess, Holland America, and Cunard. (FOE)

R. In high doses, nitrogen dioxide irritates the respiratory passages, but 40 microgram (μg) per cubic meter of air can trigger no acute symptoms whatsoever. The way the European legal threshold of 40μg came about was nothing but a bureaucratic chain reaction: way back, the EPA considered 40 microgram per cubic meter of air safe enough for a *constant indoor level* for children (thinking of gas stoves in family homes). Next, the WHO took that level as a *recommendation* for *outdoor* air. Then, the EU turned it into a *legal limit*. The USA meanwhile, introduced 100μg as an outdoor level.[167]

If a city really wants to lower its N levels anyway it should convert more factory and other flat roofs to *green roofs*. Mosses and succulents really like nitrogen as a nutrient.

S. Subsidies for fossil fuels, annually (in 2013):
- USA: US$5.1 billion in annual national subsidies to fossil fuel exploration; plus US$1.4 billion to overseas fossil fuel exploration.[168]

- UK: US$1.2 billion to oil and gas exploration, particularly shale gas and offshore resources; US$825 million to overseas fossil fuel exploration; and US$53.7 million to fossil fuel exploration projects via banks (e.g. World Bank Group, the European Investment Bank).[169]

T. Significantly, in a speech by an American president, the Inaugural Address by President Truman in January 1949.[170]

Part III

The Human Interface

Ethics and Dignity in an Era of Barbarism

We will only preserve Mother Earth through a paradigm shift from a human-centric society to an Earth-centred global ecosystem. This requires us to engage with everyone, including young people who will inherit this planet. Education is critical to safeguarding Mother Earth: training courses on harmony with nature and earth jurisprudence approaches will be essential in creating a resilient world for everyone, everywhere. I commend Member States who promote teachings from ancient cultures who have a deep connection with nature.
– Tijjani Muhammad-Bande, UN General Assembly President, 2020[1]

Chapter 15

Why Does So Little Happen, and So Slowly?

The present era, as Harvard biologist EO Wilson reminds us, will not be remembered for its wars or technological progress but as the time when men and women stood by and either passively condoned or actively supported the destruction of the living world.[1]

Humanity is totally dependent on the very ecosphere it is degrading. "We are *way past* sustainable ecological limits," say environmental scientists Haydn Washington and Helen Kopnina in a paper in *The Ecological Citizen*. "In effect, we are bankrupting nature and consuming the past, present and future of our biosphere."[2] We want so much for ourselves that we imperialistically override the needs, even the very right to live, of our fellow creatures all around the globe, and even of our own children and future generations.

Our culture seems too paralyzed to act and meet the challenge of global ecological meltdown. So much is at stake but all strata of society—politics and admin, industry, media, and consuming folks—are glued to the old paradigm of economic growth. Although every child can understand that *unlimited* growth (of economy, of human population, of consumption) is simply not possible on a *limited* planet, the almost religious belief in endless growth is still dominant. It is an illusion that reveals *magical thinking* long believed to have been overcome.

For decades the economic mantra of perpetual growth has been unrealistically held to be the cure for all of our fears and problems: poverty, unemployment, debt, destruction of nature. Only with more money acquired by "growth" would we be able to afford to solve them all, and even bring affluence and democracy to everybody in the world. But now we know that hypercapitalism and globalization actually create and amplify these problems in the first place. With all of these calamities perpetually growing, the false promise is becoming obvious by now.[A]

The world as a whole today has undoubtedly seen an incredible amount of economic growth and technological progress over the last decades, but is further away than ever from social equality.[B] "Progress" indeed brings wealth, but only to an ever-smaller minority: in 2006, the world's richest 1 percent of people owned 40 percent of all wealth; twelve years later, the richest 1 percent owned 82 percent of all wealth created. In 2014, the 85 richest people in the world were as wealthy as the poorer half of humanity; in 2018 only eight men owned as much as the poorer half.[3] It is easy to see by now how the term "to grow" itself has become perverted; "we have forgotten its original meaning: to spring up and develop to maturity."[4] In nature, growth *gives way* to a matured steady state.

Another reason why appropriate action to avoid the ecological meltdown is so hard to initiate is psychological. The American philosopher Timothy Morton coined the term *hyper objects* for phenomena which are too big for us to comprehend, or even want to look at. Capitalism is one of these, with seemingly no outside to it, so what else can there be? As the theorist Fredric Jameson said, most people find it easier to imagine the end of the world than the end of capitalism.[5] And climate breakdown is another hyper object. It is too complex, and there is no instant, or even long-term, gratification in understanding it. And we capitulate, "I am powerless, aren't I? Let *them* on the top deal with the problem. Some technology will come along to save the planet." But you are *not* powerless, nobody is. You have to start changing because nobody on the top wants to deal with the problem either. And we have the necessary technologies already at hand but are incapable of beginning to use them wisely. The hope that we don't have to change anything and that something (the "techno fix" cure-all) or someone will come along to save yesterday's comfortable lifestyle for us is pure magical thinking.

Thus the present remains "in destruction, fatalism and apathy,"[6] while the foundations of life on Earth are changing rapidly and dramatically. Without this, one cannot understand the "panicked desire to return to the former protective measures of the nation-state—which is wrongly called the 'rise of populism'," writes the French philosopher Bruno Latour in his "Terrestrial Manifesto".[7] The phenomenon of the "angry citizen" is also related to this. As so often in world history, the deep fear is directed against scapegoats. But the division into ethnic groups, into reactionaries and progressives, or into leftists and rightists make no sense anymore. The repeated tactic of playing nature conservation off against jobs and "progress" has also become obsolete.

The rise of populism and the (far) right was to be expected, the attitude of "us vs them" and "me first" is gaining ground. But owing to its lack of *real* climate action the political mainstream, however humane and politically correct its participants may deem themselves to be, is equally responsible for the "climate barbarism" that has long begun. We let our leaders attend and leave one climate conference after another without ever committing to real action while all along it is known that delay will seal the annihilation of the small Pacific islands and their cultures, as well as many coastal areas around the world, for example, Bangladesh. As Naomi Klein puts it: "A culture that places so little value on black and brown lives that it is willing to let human beings disappear beneath the waves, [...] will also be willing to let the countries where black and brown people live disappear beneath the waves, or desiccate in the arid heat."[8]

In the industrialized North, fearful citizens let these things happen because they seem far away in space and in time, and because they feel too small and insignificant to make a difference.

But global heating has begun to knock on everyone's door. The planetary boundaries are being violated by our culture, but in the end of the day nobody can argue with nature. *Nature is not negotiable.*

Fear and denial

The common fear is that with zero or negative growth the economy will stagnate and unemployment will rise. But the *idea* that economic growth and employment stand and fall together "only developed 60 years ago, and for most of human history we managed to provide employment without economic growth. [... It] is possible to develop scenarios where full employment prevails, poverty is eliminated, people have more leisure, and greenhouse gases are drastically reduced, in the context of low—and ultimately no—economic growth. It is thus mistaken to assume that economic growth is a necessity for full employment," say Haydn Washington and Helen Kopnina in a groundbreaking paper. And they warn us that "once we have exceeded ecological limits, growth will make us worse off. We have then reached *uneconomic* growth."[C,9]

But fears still linger everywhere. And as psychologists tell us, if fear stays subconscious, or if it requires action but we stay passive, all sorts of imbalanced erratic reactions can break loose. We need to face our demons!

> *The dragon is thundering behind me –*
> *his breath scorches my back.*
> *If I turn I may tame him...*
> *If I run my back will burn forever.*
> – Gabriel Millar[10]

But there is a remarkable human ability that enables us to bear long-term fear and bad conscience: *we can deny our problems!* As a society we are capable of acting "as if there is no environmental crisis, no matter what the science says."[11] This is crumbling along the edges now, but it was the dominant game for the last half century. The environmental scientist Dana Nuccitelli describes five stages in the modern global outbreak of climate denial:[12]

Stage 1: Deny the problem exists.
Stage 2: Deny we're the cause.
Stage 3: Deny it's a problem.
Stage 4: Deny we can solve it.
Stage 5: It's too late anyway.

In autumn 2018 it became dramatically clear that the Trump administration had reached stage 5. When the Environmental Impact Statement (EIS) was released to the White House, reporting that with business-as-usual the world is heading for no less than 4°C (10°F) warming, Trump had to admit to climate breakdown. But his government announced that it would still overturn Obama's CO_2 regulation program because it was "too late anyway."[D,13]

However, Trump "is a symptom, not the cause," to quote Barack Obama.[14] *Everyone* is skilled in denial. We have to look at ourselves first and foremost. Denial has been called by ecological thinkers *the single biggest reason for humanity's possible demise*. It is the greatest hurdle we are facing. And to unroot it, we have to go to the scariest place on Earth: the human psyche.

They long for painless immunity
heaven on earth
in their schizoid fortresses.
– Jay Ramsay[15]

First, let us look back at our evolution. Competition for scarce habitat and food is a selection pressure for most life forms. In humans too—homo sapiens is about 200,000 years old—this led to a tendency to occupy all accessible habitats and to use all available resources. In ecology, homo sapiens is classified as a *K-strategist*. This group tends to press up against the carrying capacities (K) of their habitats. K-strategists are usually characterized by being long-lived, slowly reproducing, extending parental care and having high rates of offspring survival. In this way they grow until their habitat pushes back—via negative feedback such as food scarcity, lack of space, disease. Despite the human history of terrible wars and huge empires, we were held in check by nature's feedback systems. Until the use of fossil fuels threw everything out of balance.

The ancient layers of survival instincts are underlying the later evolution of the human psyche. Motivational studies show that the human being is *not*—contrary to a wide-spread delusion—primarily a rational species. Emotions and instincts play a large role in human affairs. Other elements in our psychological mix are natural optimism and a tendency to favor closer people and places to distant ones, and the presence over the future. The latter phenomenon is known as *temporal discounting*. It implies that "the value of the more distant reward is diminished, or discounted, by the time intervening between the choice and the reward. Temporal discounting helps explain why it is so difficult to get teenagers to save for retirement or why it is impossible to leave a dog at home with enough food for a week and expect it to ration its consumption."[16] Or a politician to act on the climate crisis. Behavioral studies suggest that the value of future rewards declines rapidly.

Another trait of human nature is even more relevant for our understanding why societies cling to the globalized status quo. Cognitive neuroscience has shown that "repeated social, cultural, or sensory inputs can acquire a physical presence in our brains, i.e. repeated experiences and cultural norms become engrained as semi-permanent synaptic circuits."[17] We get hardwired. Such structures then act like selective filters for perception. They seek out information that confirms and reinforces our beliefs, and, conversely, deny, discredit, twist, reinterpret, or forget information that challenges our paradigms. Herein lies the root of (climate) denial.

The ecological and the ecocentrical view help us to recognize the **interconnection** of all life and the Earth systems (weather, water, forests, soil, seasons) on which life depends. The climate movements and other incentives to protect life are based on the understanding that everything is connected.

To accept that everything is connected requires us to acknowledge that our actions have consequences and therefore bring responsibilities. Something that neoliberal capitalism and its shameless exploitation of wild nature and human nature is not willing to shoulder. Conveniently, it sticks to an ideology of **separation**. So does white (male) supremacy. It is not by chance that climate denial is integral to rightwing thinking. And as social equality and ecological movements begin to support each other—the time of Divide and Rule is over—the white supremacists become increasingly nervous. "Fighting against climate change is the equivalent of fighting against hatred. A world that thrives is one where both people and planet are seen for their inextricable value and connectedness."[18]

The rising

Countless movements and events to stand up for the planet, for life, for equality and inclusion, have sprung up all around the world, and continue to grow in number and attendance. Although most people don't hear much about them—which keeps the incentives smaller than they need be. But word does spread, more via social media than the established mainstream news channels. A few examples of positive change:

- The leading author of the Hothouse Earth Papers,[19]

when asked what effective climate actions he thinks are necessary, said what all people who seriously look at climate disruption and ecosphere degradation say: the Earth has to get out of the capitalistic narrative of competition and growth, and out of the "neoliberal economy." And the speed of this has to be on a footing with that of a war economy; if humanity intends to reduce global heating to a bearable level, the industrialized North has to be as consequent as in wartime.[20]

- In October 2017, Los Angeles became the first city in the world to announce a WW2-scale climate mobilization "rooted in environmental justice", to achieve a carbon-neutral Los Angeles by 2025.[21]
- In October 2018, Molly Scott Cato MEP (Member of European Parliament) together with more than ninety academics signed a declaration[E,22] to the UK government demanding that it "tell the hard truth to its citizens", to work "in accordance with the precautionary principle", and "to urgently develop a credible plan for rapid total decarbonization of the economy." — "When a government willfully abrogates its responsibility to protect its citizens from harm and to secure the future for generations to come, it has failed in its most essential duty of stewardship. The 'social contract' has been broken, and it is therefore not only our right, but our moral duty [...] to rebel to defend life itself." Therefore, the undersigned declared their support for **Extinction Rebellion (XR)**, a project that launched on 31 October 2018.[F]

The following year saw the meteoric rise of XR activism all around the world. Full of ideas and creativity and always remaining pacifist, it quickly became a worldwide movement of peaceful civil disobedience but remained particularly strong in the UK. Apart from activist events disrupting business-as-usual at corporate headquarters, banks and political hubs (e.g. bunches of activists gluing themselves to the main entrance doors), local XR groups also offered psychological group processing of people's troubled emotions about the wholesale destruction of life on this planet. Like so many important things, XR largely disappeared from the headlines and public view when the coronavirus lockdowns were enacted.

- In December 2017, the *World Scientists' Warning to Humanity: A Second Notice* was released, carrying 15,364 scientist signatories from 184 countries.[23] As all assessments from the first warning in 1992 have manifested, or even gotten worse than predicted,[G] the renewed 2017 declaration is more urgent than ever. Kathleen Dean Moore, Oregon State University: "The next couple of years will be the most important years in the history of humankind."
- In August 2018, in the wake of Sweden's hottest summer on record, 15-year-old **Greta Thunberg** decided to go on school strike and protest outside the Swedish Parliament.[H] "I want the politicians to prioritize the climate question, focus on the climate and treat it like a crisis,"[24] she said. After two lonesome weeks on the cobblestone, more and more people began to join her protest in September. In December, she was invited to speak at the UN climate conference (a must-see: "Greta Thunberg's COP24 speech in Katowice 2018" on YouTube).[25]
- On Friday, 30 November 2018, thousands of school children across Australia followed the "Strike 4 Climate Action" school strikes, which were then repeated every Friday until the end of the COP24 climate conference in Poland. The inspiration was Greta Thunberg in far-away Sweden, the trigger was an extraordinarily disturbing heatwave. In the days running up to the first school strike, Australia's prime minister and the resources minister[I] disapproved of the civil disobedience and mocked the

children, but got their own back. In fact, it only inspired the most apt and witty demo posters and banners, for example, "Why study science, our governments have stopped listening," "I've seen smarter cabinets at Ikea," or "Denyosaurus – fossil fool."[26]

Interview comments were equally refreshing. Manjot Kaur, 17, said: "If [prime minister] Scott Morrison wants children to stop acting like a parliament, then maybe the parliament should stop acting like children."[27] By 4 December, more than 20,000 students had joined school strikes in at least 270 towns and cities in countries around the world.[28] The week of 15 March 2019 saw 1.5 million strikers in about 2,000 places; and on 24 May, more than 1.8 million people joined the climate strikes in 2,350 cities across 125 countries.[29] **#FridaysForFuture** school strikes, climate demos and Extinction Rebellion turned into massive and fast-growing international movements. Established parties and fossil-fool politicians around the world found themselves dumbfounded and outwitted by the sudden political engagement and eloquence of the younger generations.

During the following nine months the global climate movement inspired by Greta Thunberg went from strength to strength. Millions of young people, and from autumn 2019 onward older generations, too, joined the protest marches of Fridays For Future. The impulse grew ever stronger until street protests and public gatherings were brought down to a halt by the COVID-19 lockdowns which began in March 2020. Historically, the "social distancing" represents the climax of our society's underlying obsession with separation. It is also typical for a society alienated from nature that—although microorganisms are the very matrix of life—we choose to wage war on "evil" bugs instead of asking why our immune systems have become so run-down in the first place.

- Yet the movements for climate protection, and to prevent further biodiversity loss, both suppressed by corona lockdowns, also benefit human health. It is, after all, the toxins from burning fossil fuels that are among the leading causes of illnesses and premature deaths in the world (in 2018, air pollution killed 8.7 million people worldwide[30]).[J] *Climate protection and ecosphere protection equals health protection.*

What can I do?

- Follow your conscience uncompromisingly. Listen to your heart. Think of all the children and teenagers of human and all other species who deserve a healthy life on a healthy planet. What can you do *for them?*
- A conscious personal lifestyle is not everything. In 2020, global corona lockdowns reduced greenhouse gas emissions by only 4.5 percent (CO_2 by 7 percent)! *We as citizens must also demand far-reaching systemic changes on the political and industrial levels.*
- If you feel a bit unsure about acts of civil disobedience, think of the words of the philosopher Juergen Habermas: "Every constitutional democracy which is sure of itself views civil disobedience as a necessary ingredient of its political culture."[31]

For the last fifty years we have looked the other way while the branch on which we sit was being cut. Now civilization is hanging on a mere silken thread. If we only let her Mother Nature could weave new ones, until she has made the web of life strong again. And we may remain a part of it, if we finally find our balance soon within the "grand organic whole".

Footnotes

A. Its absurdity has long been foretold by the political scientist André Gorz. He pointed out as early as 1974 that "the idea that growth reduces inequality is a faulty one."[32]

B. Huge new middle classes have grown in the "threshold countries" which seems to contradict this statement, but this is outweighed by the global disparity of astronomical super-wealth centered in the "first world" versus billions of people (mostly in the southern hemisphere) becoming ever poorer.

Even the International Monetary Fund (IMF) now recognizes that the globalization *crisis* is a social crisis. The IMF's 2018 program represents a total reversal of the past. The new keyword is "inclusive growth". It aims to promote economic growth "by also promoting social justice, by expanding social networks. A strategy paper of the International Monetary Fund states that infrastructure, education, childcare, increasing women's labor force participation and much more should be promoted," says economic expert Rudolf Hickel.[33]

C. Detailed scenarios of alternative forms of civilization are outside of the scope of this book, but see Chapters 17 and 18 for a start. Alternative concepts for the economy can be found in Ernst Ulrich von Weizsaecker, Anders Wijkman, et al. 2018. *Come On! Capitalism, Short-termism, Population and the Destruction of the Planet. A Report to the Club of Rome*. New York: Springer.

D. The proposal of the Trump administration stated that the US transport sector alone won't make enough difference to global CO_2 emission, and that the climate action would also "require drastic reductions in all U.S. sectors and from the rest of the developed and developing world," which would imply the US "economy and the vehicle fleet to substantially move away from the use of fossil fuels, which is not currently technologically feasible or economically practicable."

Pathologically Denial Stage 5![34]

E. The declaration says that running a totally unsustainable form of civilization with no knowledge about the sensitivity of global tipping points amounts to a blindfold global experiment, and that hence "our government is complicit in ignoring the precautionary principle, and in failing to acknowledge that infinite economic growth on a planet with finite resources is non-viable. Instead, the government irresponsibly promotes rampant consumerism and free-market fundamentalism, and allows greenhouse gas emissions to rise."[35]

F. *The Guardian* columnist George Monbiot also joined Extinction Rebellion and said in an interview: "The social contract has been broken [...] the government is not looking after us! What we're facing is a complete institutional failure to respond to the existential crisis that we're in. We are literally losing our life support systems! Climate breakdown, environmental breakdown in general, [is] driven by this corporate capitalist model which is simply not suited to the preservation of the world's living systems." He concludes that this crisis requires a response of civil disobedience.[36]

G. The date commemorated the first *World Scientists' Warning to Humanity in 1992* which was signed by 1,700 independent scientists, including the majority of living Nobel laureates in the sciences. This first warning had already urged humanity to alter its collision course with the natural world, and warned about all aspects of ecosphere destruction and climate change, as well as continued human population growth. "The authors of the 1992 declaration feared that humanity was pushing Earth's ecosystems beyond their capacities to support the web of life. They described how we are fast approaching many of the limits of what the ecosphere can tolerate without substantial and irreversible harm."

Recommended: ScientistsWarning.org.

http://www.scientistswarning.org

Also: Oregon State University video clip about *The Second Warning*.[37]

H. In June 2017, Sweden had enacted the world's then most ambitious climate law—attempting to get GHG emissions to zero by 2045—but "this is too little too late, it needs to come much faster," says Greta. "Sweden is not a green paradise, it has one of the biggest carbon footprints."

I recommend Greta Thunberg's article: Greta Thunberg 2018. Sweden is not a Role Model. *Medium*, August 24.[38]

I. Prime minister Scott Morrison, resources minister Matt Canavan.[39]

J. Soot and particulate matter from car and industrial exhaust settle in the trachea, bronchial tubes and lungs of humans and animals, where they can trigger all sorts of infections. By destroying certain immune cells, particulate matter weakens not only the immune system, but also DNA repair, cell regeneration, and the cardiovascular system. The predisposition to cancer is also significantly triggered by air pollution, pesticides and heavy metals (especially from brake disc abrasion).[40]

Chapter 16

Anthropocentrism

Human-centredness or anthropocentrism (from Greek *anthropos,* "human being") sees humanity as central to the universe. The belief that the human species is the most important entity in the cosmos has also provided an all-too-firm base for concepts of human supremacy. Anthropocentrism interprets the world in terms of human values and experiences. It results in the degradation of the entire ecosphere of the planet which loses its inherent value, beauty, dignity, and sacredness, only in order to provide raw materials for human production and consumption. Being so important and central, humanity sees itself as separate from the "rest" of the Earth, and calls it *environment,* meaning "that what surrounds, encircles"—a term which always points back to the central supreme being called "Man" (indeed, patriarchism has been very entangled with the history of anthropocentrism).

There is wide-spread agreement among ecological thinkers and deep green philosophers that **anthropocentrism is the root cause of the global ecological meltdown**.

Breaking free from anthropocentrism has the potential to unite social and environmental groups and movements. Both ecosphere degradation and social inequalities based on gender, skin color, age, disability, caste and class are based on the underlying acceptance of the unjust consumption, abuse and depletion of the life force (or even the life) of other beings for personal gain.

Anthropocentrism comes in two main forms: religious and secular. In religious anthropocentrism, "the universe exists solely as a stage on which the drama of humanity's fall and redemption is played out, and once that drama is over, God can be expected to haul away the universe as we know it and replace it with one that humanity likes better," as the author JM Greer sums up so aptly.[1] In secular anthropocentrism, to the believer in progress, "the universe exists solely as a stage on which the very different drama of humanity's conquest of nature is to be enacted, and the replacement of the universe as we know it with one more subservient to our whims is supposed to be accomplished by science and technology." However, this would require that the limited mental capacity of our species would be sufficient to *understand* the nature of the universe, and this in itself "was never more than an act of faith". At the end of the day, the faith in progress is an ersatz religion that *promises salvation* as well, just in a different way.

For all their destructiveness, both forms of anthropocentrism nevertheless had a life-affirming aspect once. When secular anthropocentrism was born in the age of Enlightenment, it was a creative way out of the horrendous religious wars between Catholicism and Reformism, imagining itself to be lifting up, without disparity, "every human being from the caves to the stars."[2] And religious anthropocentrism, in its dawn in antiquity, dispersed the merely ethnocentric mindset (the "us vs them" mentality of single ethnic groups) and created common ground with other members of the human species. Still, ethnocentrism ("tribal consciousness") has never fully disappeared, and today breaks out as primitive and crude as ever in all corners of the world.[A]

Furthermore, the anthropocentric age is characterized by intense hierarchism: we might be all human, but white is better than dark, male better than female, straight better than gay. All these forms of mental sickness have never disappeared. But one thing is for sure: they are all completely outdated now.

It is high time for another historic leap in consciousness. Into a new way of understanding our presence on Earth as being a part of the whole. Homo sapiens is getting ready to take the step "from conqueror of the land-community to plain member and citizen of it," as Aldo Leopold said.[3] All beings—human and non-human!—are Earth citizens, with equal rights to share the planet and live a fulfilled, happy life. This view of existence is called *ecocentrism* (see next chapter). The shift from human-centredness to Earth-centredness triggers fearful resistance, of course. But as Stan Rowe, the great pioneer ecological thinker, reminds us: "A few hundred years ago, with some reluctance, Western people admitted that the planets, sun and stars did not circle around their abode."[4] Letting go of anthropocentrism just means kissing goodbye another layer of human hubris and self-deluded grandeur.

And this shift of perception and heart is very urgent, not just because of global climate and ecosphere breakdown. For one, none of the "environmental" efforts of the last 50 years has been able to change the system or its stubborn trajectory into mass extinction. Everything remained piecemeal first aid to local symptoms, or confined to do the monitoring and bookkeeping of the great destruction. And why? Because most environmental efforts, even with the best of intentions, are still steeped in anthropocentrism. And still the majority of environmental institutions and organizations understand conservation predominantly as a service to *us* instead of the ecosphere as a whole. Even the UN Sustainable Development Goals from 2016 are fundamentally human-centered rather than Earth-centered. In recent years, the development of the concept of "ecosystem services" still worships this old paradigm: everything in nature has a price label—according to what it is worth *to us*, and can be traded like carbon permits.[5]

And secondly, it is urgent because the current system is totally unethical, brutal, murderous, "morally bankrupt,"[B,6] and should not be sustained a minute longer. To get the picture we need to face the dragon one more time, and look at the dark side of our "civilization", the one we never wanted to see. But meeting our shadow makes us more humble, and at the same time gives us an entirely new sense of strength, one that comes from authenticity and honesty. We need all these qualities if we want to contribute to a good future for the whole planet.

Wétiko

The Native American scholar and activist Jack D. Forbes[C] published a radical critique of Western civilization in 1979 which became a founding text for the American Indian Movement (AIM). His history of terrorism, genocide and ecocide was provocatively named *Columbus and Other Cannibals*.[7] When Christopher Columbus reached the West Indies he found a people totally different from Europeans. Indeed they were so peaceful, trusting and welcoming that the Spanish named them Indios, *in Dios*, "in God". Columbus sang their praises too, in a letter to the Spanish Crown: "Of anything they have, if it be asked for, they never say no, but do rather invite the person to accept it, and show as much lovingness as though they would give their hearts… And they know no sect nor idolatry; save that they all believe that power and goodness are in the sky…" But in the *same* letter he continues: "As soon as I arrived in the Indies […] I took some of them by force… Their [Spanish] Highnesses may see that I shall give them [the Spanish Crown]… slaves as many as they shall order to be shipped."[8] Columbus then proceeded to ship thousands of the "loving" people to Europe and Africa for profit. This schizophrenic slave trader and big-style human trafficker we call Columbus was only the beginning. Over the following two hundred years, the Americas would lose over 90 percent of their population, that is *130 million human beings*, to slavery, genocide, and imported diseases.[9]

What Forbes' work unravels, however, is the *mindset* of the "white man" that enabled him to continue on a path of extraordinary brutality and unspeakable atrocities; and how this mindset of betrayal, aggression, destruction, exploitation and greed corrupts many that get in touch with it, also among the victims. Forbes historically traces back this grave mental disturbance through European and Asian history, and to the first evidence in ancient Egypt and Mesopotamia. And he calls it *wétiko,* after a Cree word for "cannibal". It does not denote the kind of traditional ritualistic cannibalism as an act of eating a small portion of a dead enemy's flesh as an act of honoring this person. But "an evil person or spirit who terrorizes other creatures by means of terrible evil acts". Forbes says, "*Cannibalism,* as I define it, *is the consuming of another's life for one's own private purpose or profit.*" And he warns that "this disease, this *wétiko* (cannibal) psychosis, is the greatest epidemic sickness known to man."

Tragically, nothing of this is only history. When in early 2019, the Brazilian president announced that he will loosen regulations for conservation, and also for firearms so that miners and adventurers can subdue the rainforest and take land from the Indios even in their reservations, he was in all clarity announcing the beginnings of a genocide. And how did the "democratic" countries react? They signed a trade deal with Brazil a few months later (Mercosur). Can we resign from "the insane trading of rainforests for crop plantations and cattle ranches"?[10] This is *wétiko* cannibalism, the consumption of the life of others for one's own purpose or profit. And it is everywhere: the slave-like conditions for laborers in mines, electronics and clothes factories, etc., etc. Land grabbing, the destruction of habitat, factory farming and livestock death camps. The *extraction economy,* at its ice-cold heart, is a cannibal culture. And we are not able to transform this into something ethically just and graceful until we part from anthropocentrism. Let us dismantle some of the predominant lies of the anthropocentric narrative.

The first lie: "Everything is war"

We have been told that everything in nature is at war, that evolution favors the strong and crushes the weak. This is a crude misinterpretation of nature, *and* of Darwin's work, respectively. Charles Darwin (1809-1882) lived during Britain's imperial century, and hence in an epicenter of the pinnacle of man's hubris as colonizer of the world, but nevertheless he managed to lay foundations for a deeper understanding of the Earth. But it was inevitable that Darwin's ideas got hijacked by the warmongers. "Survival of the fittest" is a term not even coined by Darwin himself but by the sociologist Herbert Spencer. Ever since it has fed notions of inequality and superiority: "the fittest win", losers are weak and pathetic, and the "rest" of the planet is for the taking. Darwin hesitated so long to publish his work on evolution[11] not only because he was sensitive to the Christian faith but also because he feared this other danger of socio-political abuse.[12]

In the second edition of his book on evolution, Darwin adapted the term "survival of the fittest", but for him evolution was about the *long-term development of species*. Evolution is not about *individuals fighting* each other but about how species develop in and with their environment over tens of thousands of years. It's never cruel or "inhumane" to any individual organism; they don't even *know* if their species is fading out and being surpassed by another. And since Darwin, genetics have shown that *there never were winners or losers*: the DNA of the not-so-favored characteristics doesn't disappear, it just goes dormant (some geneticists have shortsightedly called it "junk DNA"). Hence, the "winner" is not a prize for eternity, just the *temporary response* to certain conditions surrounding the organism. On this level, there is no winning and losing in nature. But worryingly, the GenDrive technology (see Chapter 10) is set up to manipulate nature in this respect and by destroying the "junk DNA" really create the competition that never was there before.

Take the co-evolution of a plant and an animal species that likes to eat that plant. The plant develops a poison to stop the feeding after a few leaves. Over time, the animal develops some resistance, and the plant changes the recipe for the poison. The animal adapts again and the plant reacts in turn. This kind of process is usually and unquestioningly *interpreted* as a "battle for survival". I could just as well call it a dance. Because nature is abundantly inventive and lets all sorts of possibilities evolve and play out. But immediately I would be ridiculed as a romantic and an idealist. Maybe I am, but I stay put: "battle" too is merely an interpretation.

Industrial society is completely steeped in the narrative of war. Military jargon is omnipresent in medicine and biology (colonization, defense line, outbreak, attack, antigen, antibody, antibiotic); agriculture—whereas growing food should be an art form that works in harmony with nature—is another hotspot for battle, with a wide arsenal of chemical weapons and mechanical slaughter devices; and social and economic life is rife with competition, because "only the fittest win". Thus, the war paradigm justifies all forms of cannibalism, of eating the Earth, and each other. Eat, so you don't get eaten! The basic mode of modern life is fear. Capitalism cunningly plays on ancient (Ice Age) anxieties. In a nutshell: Go shopping now, or you either starve or get eaten by a big saber-toothed cat!

In truth, not even that horror is a given; getting eaten by a big cat or wolf might not be as abysmal as humanity had thought for so long. The brain of the deer or antelope that is running for its life from an approaching predator is flooded by adrenalin and panic hormones. At some point during the chase, the emotions turn from panic to ecstasy. At the first strike of the predator, the deer goes down and into "freeze" mode, its state of consciousness changes. In the face of death, there is only acceptance and subjection. The body floods with endorphins, the body's own painkillers (chemically related to morphine), so when the predator bites the deer won't even feel pain. The international expert on trauma healing, Dr. Peter A. Levine, has studied trauma for over thirty years. He quotes the report by the explorer David Livingstone who had been saved at the last minute from a lion's hungry stomach. The animal shook him and that induced a state of mental surrender and physiological anesthesia. (In psychology, the remainder of this state of "dissociation" is key to the trauma healing in survivors.)[13] In the moments before death there is a state of heightened consciousness without the limiting fears of the ego. I claim that it is *this surrendering* which the Western mind fears more than actual death. But this is the price for being anthropo- and egocentric. (Personally, surrendering to an animal predator seems more humane and enlightening to me than a slow death in hospital.)

If anyone still claims "Nature is war" I would like to say this: "You enjoy a body consisting of about a thousand billion cells, each of which can only exist because it lives in symbiosis with tiny mitochondria inside that are aliens with an entirely independent DNA to yours. Your cells form complex structures of organs and exchange systems which all work together to maintain the vitality of your body. And your cells are aided by ten times their number of bacteria which cooperate within your body's "superorganism". At least five hundred species among them could kill you within a day if they went berserk, but they don't. In your body alone, there are more acts of cooperation every second than you have days in your life. And you want to tell me that everything in Nature is war?!?"

The second lie: "The march of progress"

We have been told that history was a continuing rise of human society to ever-new heights of brilliance, intelligence, and subduing the Earth. The step from gathering and hunting to agriculture supposedly was one of the most ingenious undertakings of our forebears, and via further enhancing our tools and methods we got to the Industrial Age and the Digital Revolution. We are the "crown of creation". Those Indigenous people who still exist scattered around the planet hardly even stepped out of the Stone Age, that's why "superior" Europeans used to call them "primitives" and "savages". But in truth: was the step to agriculture really such a glorious one?

Surprising insights from more recent historical, archaeological and genetical research force us to paint the picture anew. To start with, agriculture and the sedentary lifestyle did not present an Edison-light-bulb moment to everyone who saw it; people did not immediately flock to it. On the contrary, it took more than four millennia (!) from the first crop domestications and the first sedentary (fixed residence) settlements to the early small, stratified, tax-collecting, walled city states of Mesopotamia. Nobody was keen on entering those hot-spot melting pots of people, livestock, bad sanitation, and germs. After all, with homo sapiens about 200,000 years old, our species spent about 95 percent of this time being nomadic gatherers and hunters. The safety for them lay precisely in their mobility and the diversity of food sources which were abundant, stable and resilient: mass migration of big game and of birds, abundance of small game, waterfowl, fish and other aquatic creatures, plus all the abundance of plants from diverse ecosystems (in the Mesopotamian plain reaching from wetlands and floodplains to savannah and woodlands).

These gatherers and hunters had a lot of spare time for socializing and other pleasant activities.[D] What the nomads saw when they visited the cities was mainly drudgery. Agriculture is immensely labor-intense. In the Neolithic period, not only plants and livestock were domesticated, but also homo sapiens, who now was tied to the round of ploughing, planting, weeding, reaping, threshing, and grinding, ceaselessly tending to the daily needs of their grains and livestock. Hence the early pastoralists and hunter-gatherers, as Yale anthropologist James C. Scott says, "fought against permanent settlement, associating it, often correctly, with disease and state control."[14]

By no means were these early towns and cities paragons of civil peace, social order, and freedom from fear that drew people in by their charisma and splendor. The unprecedented concentration of human and livestock numbers also attracted uninvited guests like rats, mice, weevils, ticks, and inspired many pathogens to spread and crossbreed. The results are still known today: whooping cough, meningitis, diphtheria, polio, smallpox, measles, mumps, infectious hepatitis, and many other diseases began their career with the agrarian revolution. Their rise was further assisted by the rather poor diet of the farming population which was much less varied than that of the hunter-gatherers.[E] Interestingly, the choice of the very few domesticated plants—predominantly wheat, barley, rice, chickpeas, and lentils—had *nothing* to do with their dietary value or food safety (resilience to weather changes, etc.) but with something entirely different: taxation.

Grains are the best suited for tax assessment, appropriation, cadastral registration, storage and rationing. Other favorite domestic food plants of the time, like manioc and yucca, require little care, also ripen within a year, can be left safely underground for two more years—but also can be hidden from the tax collector. Even domesticated legumes, such as peas, soybeans and peanuts, don't serve so well as tax crops because they are "indeterminate" crops, i.e. they can be picked as long as they grow. Wheat or rice fields are the perfect stuff for state-building.

State formation depends on control, maintenance, and expansion of the concentrations of grain and manpower. Because of recurring severe population losses (due to wars or pandemics) the city-states made extensive use of unfree labor and slavery.[F] Many "barbarians" preferred to stay outside the domesticated area; "the line on the frontier where the barbarians begin is that line where taxes and grains end," says Scott.[15] And the cities had other problems: the population density caused steady deforestation of the upstream waterways which led to siltation of the riverbeds and floods; and intensive irrigated agriculture resulted in salinization of the soil, lower yields, and eventual abandonment of arable land. Such ecosystem degradations led to collapse of some cities and migration of its inhabitants. The ecological price for grains and grazing is loss of soil fertility, soil erosion, and landscape degradation, and it continuously pushed human expansion.

Over time, the city populations developed some immunity against the urban storms of parasites, and another biological effect came into play. With a diet high in carbohydrates, puberty arrived earlier, menopause later, and ovulation became more regular. Because farming communities always needed more labor for the fields, they did not, as the "barbarian" gatherers and hunters did, practice delayed weaning or contraception. Despite terribly high infant mortality rates, and adults generally in poor health,[16] the agrarian populations began to overshoot. The economic order that derived from agriculture drove human expansionism, and populations began to outgrow the carrying capacity of their homelands. But we did not inherit a tale of early ecological warning;[G] what we inherited is the narrative of the "ascent of man", of the superiority of sedentary fixed-field farming, of how agriculture replaced "the savage, wild, primitive, lawless and violent world of hunter-gatherers and nomads."[17] The belief in "progress" and the "march of civilization" has continued ever since.

It is here, in early agriculture, where the crucial split in the human psyche took shape. The agricultural field has such a high degree of organization and discipline which is constantly threatened by "wild" nature surrounding it: weeds "invading" the field, insects or birds eating "our" seeds, wind and weather adversities threatening the harvest. Wilderness became "the other" side, the uncontrollable, dark, evil, fearsome force of nature. These are the roots of anthropocentrism, about ten thousand years old. "The relationships among humans and between humans and the more-than-human world were fundamentally altered with agriculture, while the material reproduction of humans became a self-referential enterprise with the integrity and force of a 'superorganism'," says Lisi Krall, Professor of Economics at New York University. Agriculture was the result of a "complex play of evolution on external and internal factors that set the course and stage for the rise of global capitalism and the sixth mass extinction."[18]

The third lie: "The pinnacle of intelligence"

We have been told that no other creature is as intelligent as the human being. In our arrogance, we deem ourselves far above the other animals, we "patronize them for their incompleteness, for their tragic fate of having taken form so far below ourselves. And therein we err, and greatly err," wrote the naturalist Henry Beston as early as 1928. "For the animal shall not be measured by man. In a world older and more complete than ours they move finished and complete, gifted with extensions of the senses we have lost or never attained, living by voices we shall never hear. They are not brethren, they are not underlings; they are other nations, caught with ourselves in the net of life and time..."[19]

Not even agriculture and the division of labor are inventions of homo sapiens, but of ants and termites who practiced these long before any humanoids did. Leafcutter ants (*Atta*) process their cut leaves in an assembly line that involves a complex division of labor. They *grow* their food.[H]

Most people, however, have more direct experience with the intelligence of cats, dogs, or horses than with ants. Or have seen some remarkable video clips of crows solving labyrinth and other riddles. However, the most enlightening examples of more-than-human intelligence are the cetaceans––whales and dolphins. Captain Paul Watson, a Canadian veteran marine conservation activist, says that for decades, cetologists have observed, documented and deciphered the vast intelligence of whales and dolphins. An intelligence that predates the evolution of human primates by millions of years. The near future may prove their associative, linguistic and cognitive skills to be far superior to those of humans, which is probably why we don't hear much about it. It may just challenge human pride too much, or we lack the means to understand it anyway. As the physician and neuroscientist Dr. John C. Lilly (1915-2001) said: "What I found after twelve years of work with dolphins is that the limits are not in them, the limits are in us."[20]

"Dolphins and whales do not display intelligence in a fashion recognizable to [our] conditioned perception of what intelligence is, and thus for the most part, we are blind to a broader definition of what intelligence can be," says Captain Watson.[21] "Humans evolved as tool-makers, obsessed with danger and group aggression. This makes it very difficult for us to comprehend intelligent non-manipulative beings whose evolutionary history featured ample food supplies and an absence of fear from external dangers." Watson, who has observed whales and dolphins in the wild for fifty years, describes their "discriminatory behavior in their dealings with us, treating us not like seals fit for prey but as curious objects to be observed and treated with caution. They can see beyond the manifest technological power that we have harnessed, and they can adjust their behavior accordingly. It is a fact that there has never been a documented attack by a wild orca on a human being. Perhaps they like us. More likely they know what we are."

The mental capacities of cetaceans put into question our very way of defining and measuring intelligence. For us, says Watson, "technology automatically indicates intelligence," and its absence "translates into an absence of intelligence." We are so engrossed in our limited view of intelligence, measuring it in strictly human terms, that in order to acknowledge a superior one, it would have to step out of a spaceship holding a sophisticated laser gun. For us, "intelligence is not a naked creature swimming freely, eating fish, and singing in the sea."

Yet, a whale is an organic submarine, and cetaceans have *organically* developed their sonar for hunting and communication. "Imagine being able to see into another person's body, being able to see the flow of blood, the workings of the organs, and the flow of air into the lungs. Cetaceans can do this through echo-location. [...] If an animal is drowning, this becomes instantly recognizable from being able to 'see' the water filling the lungs. Even more amazing is that emotional states can be instantly detected. These are species incapable of deception, whose emotional states are open books to each other. Such biologically enforced honesty would have radically different social consequences from our own." What's more, since sonar transfers information as an audible signal, it is likely that cetaceans can communicate such holographic "images" via a single songline. While our languages are analogue, whales and dolphins can communicate digitally.

Biology has taught us that the human brain is the most developed of all mammalian brains, and the most complex in organization and structure. This isn't quite true. The cetacean brain is uniquely different in its physiology, and more evolved: it has a fourth lobe while our brain only has three.[I] We cannot fathom what state of consciousness dolphins and whales have or what it "feels" like to be one, but anatomy and behavioral science give us some clues.[J] For example, primary sensory processing relative to problem-solving is a significant indicator for intelligence. This "associative ability"—the connecting of ideas—is a measurable skill. A rat's associative skill is measured at 9:1, meaning that 90 percent of its brain power is engaged with sensory processing, leaving only 10 percent for associative skills. A cat has 1:1, i.e. half the brain is available for association. A chimpanzee has 1:3, and a human being 1:9, meaning that we only need a tenth of our brain power to operate our sensory organs. Whales and dolphins, because the fourth lobe in their brains frees up a lot of brain power, have 1:25 on average, and can range up to 1:40.

If you prefer the more familiar human Intelligence Quotient (IQ): based strictly on morphology and cortical structural development alone, a dog scores about 15, a chimpanzee around 35, a human 100, and a sperm whale 2,000.

"Humans may be the paramount tool-makers of the Earth," concludes Watson, "but the whale may be our paramount thinker." And what is intelligence anyway? It evolves in a unique way within each species to help it along. Could you spin a spider's web or make a bird's nest with only a beak instead of hands? A bee, a shark, a rose, or a panther are better than anyone in being a bee, a shark, a rose, or a panther. "A complex intelligence exists within every sentient creature relevant to its needs." Whales don't need cars or rockets, "perhaps they have already discovered that the ultimate destination of a voyager is to arrive back where it belongs—in its own place within the universe." Which for the cetaceans and us… is Earth.

What can I do?

- Recognizing the advanced evolution of whales and dolphins has profound moral implications. How can we stay silent about them being caught, harmed or killed as bycatch, or tortured by military and oil-digging sonar? You can check out Greenpeace's ocean program, the WWF, and sign petitions when they come your way.
- Regarding anthropocentrism and its narratives: Always be alert and question habitual assumptions, thought

patterns and dogmas.

- Be watchful of concepts of "the enemy". If we are trying to create a more just world, "we cannot continue to have ideas of the 'other' that needs to be conquered. That's conquest mentality that led us to the place that we are in today. We've got to shift that way of thinking." (Sherri Mitchell)[22]
- Avoid finger-pointing. Are you greener than thy neighbor? Let us stop the squabbling. Letting go of habits, and the task of down-sizing is hard for everyone. We all stumble, fail, and get up again. Let's focus on the real task of ecosphere protection.

Footnotes

A. Fundamentalism, Islamism, fascism, nationalism, populism, etc.

B. "But the idea of commodifying nature for economic and population growth is morally bankrupt. It seeks only to legitimize human manipulation and exploitation and ultimately is a threat even to human survival." (George Wuerthner)[23]

C. Jack D. Forbes (1934-2011) was a writer, scholar and political activist in Native American issues. He was co-founder of the American Indian Movement in 1961 and started a program in Native American studies at the University of California, Davis in 1969, and was one of the founders of the first tribal college in the USA, the Deganawidah-Quetzalcoatl University, in 1971.

D. For example, still today, from large stands of *wild* wheat in Anatolia "one could gather enough grain with a flint sickle in three weeks to feed a family for a year," says James C. Scott in his landmark book *Against the Grain: A Deep History of the Earliest States* (Yale University Press).[24]

E. Agriculturalists lived off a few dozen plant and animal

species, but at a single hunter-gatherer site remains of 192 different plants were found, of which 118 are known to have been eaten by humans.

F. City population losses were not only caused by war, invasions, and pandemics but also by the general incredible burdens for non-elite citizens. Hence the extensive use of unfree labor comprised war captives, indentured servitude, slavery, forced resettlement in labor colonies, convict labor, and communal slavery.

G. Although the *Epic of Gilgamesh*, the oldest surviving tale of humanity, is exactly that: a dire warning of male hubris and ecological abuse, leading the hero into the darkest abyss. Known in ancient Greece, where it substantially influenced the *Iliad* and the *Odyssey*, it completely disappeared during the successive ages. Only to resurface in 1880 with the first translation of Assyrian cuneiform tablets by George Smith.

H. The leaves are cut to ever smaller pieces, formed into pellets, and planted with loose strands of fungus, which is their staple diet.[25]

I. While the human brain consists of three segments—the rhinic, limbic, and supralimbic—with the neocortex (the bit with all the convolutions, looking like a walnut) sitting on top of them all, the cetacean brain is uniquely different: here, "we see a radical evolutionary jump with the inclusion of a fourth segment. [...] No other species has ever had four separate cortical lobes."

J. For inter-species measurements of intelligence, more important factors than brain size are neural connectivity and complexity, sectional specialization, and internal structure, also the extent of lamination, the total cortical area, and the number and depth of neocortex convolutions, as Captain Watson explains.

Chapter 17

The Ecocentric Worldview

Earth-centredness or ecocentrism (from Greek *oikos*, "house", our home being the ecosphere of this planet) finds intrinsic (inherent) value in all of nature and the ecosphere. It sees the Earth as central to life. In the words of Stan Rowe: "All organisms are evolved from Earth, sustained by Earth. Thus Earth, not organism, is the metaphor for Life. Earth not humanity is the Life-center, the creativity-center. Earth is the whole of which we are subservient parts."[1] Not denying "the undoubted importance of the human part", ecocentrism acknowledges that "the whole ecosphere is even more significant and consequential: more inclusive, more complex, more integrated, more creative, more beautiful, more mysterious."[2]

Ecocentrism recognizes that humans have responsibility towards the ecosphere, moral sentiments that are increasingly expressed in the language of rights. Such "rights of nature" are now becoming enshrined in a slowly growing number of national constitutions, and are variously termed Earth jurisprudence, rights of nature, or Earth Law (see next chapter).

In the ecocentric view the inherent value of a creature or ecosystem is not based on its performance to serve other members or a bigger whole. They don't have to prove their worth in terms of "usefulness"; they have a right to be, simply because they exist.

Ecocentrism has some roots in Deep Ecology which developed in the 1970s. But while Deep Ecology honors the whole of nature as a metaphysical unity, ecocentrism values differences and emphasizes alliances and solidarity between them.[3] Ecocentrism does harmonize with biocentrism and zoocentrism but is more encompassing: while zoocentrism sees inherent value only in animals, and biocentrism in all living things, the broader term ecocentrism is the most inclusive concept, acknowledging that the ecosphere is the primary life-giving matrix that sustains all. Thereby ecocentrism is incorporating the Gaian perspective that the biotic and abiotic elements together form the "grand organic whole".

We are part of this whole, and always have been. Humans evolved with all other species out of the ecosphere's rich web of life. Other species literally are our relatives, close or distant. This biological kinship calls us to respect the rights of every being to exist, thrive, and evolve on its own terms. All life is interdependent, and both human and more-than-human organisms are absolutely dependent on the ecosystem processes that nature provides. Human subsystems, like economy, are also dependent on the ecosphere, and must treat it accordingly instead of degrading it. As the *Statement of Commitment to Ecocentrism* says: "Ecology teaches humility, as we do not know everything about the world's ecosystems, and never will. This leads quite naturally to a precautionary approach towards all the systems that constitute the ecosphere, so that where there are threats of serious or irreversible damage, a lack of full scientific certainty ought not to be used as a reason for postponing remedial action."

In *A Manifesto for Earth*,[4] two ecocentric pioneers, Ted Mosquin and Stan Rowe, have outlined the core principles of ecocentrism:

Core Principles of Ecocentrism

1. The ecosphere is the center of value for humanity.
2. The creativity and productivity of Earth's ecosystems depend on their integrity.
3. The Earth-centered worldview is supported by natural history.

4 Ecocentric ethics are grounded in awareness of our place in nature.
5 An ecocentric worldview values diversity of ecosystems and cultures.
6 Ecocentric ethics support social justice.

Action principles
7 Defend and preserve Earth's creative potential.
8 Reduce human population size.
9 Reduce human consumption of Earth parts.
10 Promote ecocentric governance.
11 Spread the message.

To avoid a common misunderstanding: ecocentrism is not anti-human, though it rejects chauvinistic anthropocentrism. It promotes "a quest for abiding values—a culture of compliance and symbiosis with this lone living planet—it fosters a unifying outlook."[5] The opposite perspective, looking "inward" (into human affairs solely) without comprehension of the "outward" (the ecosphere), keeps us entangled in warring humanistic ideologies, religions, sects, and tribal, racial, or national groups and their multiple squabbles with each other. Spreading the ecocentric message, emphasizing humanity's shared outer reality, offers the first trustworthy foundation for *real* conservation work, and "opens a new and promising path toward international understanding, cooperation, stability, and peace."

What can I do?

- Learn more about ecocentrism on *The Ecological Citizen*: ecologicalcitizen.net

- Support the budding movement of ecocentrism. For example, by signing the above-mentioned *Statement of Ecocentrism*: www.ecologicalcitizen.net/statement-of-ecocentrism.php
- Spread the word.
- Never forget that, despite the term, ecocentrism is not an -ism, not a new ideology, not seeking to exclude or to be superior. The task at hand is to be *inclusive,* to unite and find solidarity with everyone who is ready to commit to Life. The restoration of Earth's ecosystems, the overcoming of social discrimination (racism, chauvinism, ableism, etc.) and animal cruelty are all necessary steps towards a healthy planet.

Chapter 18

A Budding Future

Earth Law

All beings depend on the Earth and her life support systems, so it is obvious that the first and foremost laws of human society should acknowledge and protect the very foundation on which human culture and well-being is based. In all the original and Indigenous cultures around the world, "natural laws set the foundation of a respectful and harmonious conviviality among humans and non-humans."[1] The Great Law of Peace of the Haudenosaunee (Iroquois) Confederacy,[2] for example, incorporated the seven-generations-principle, meaning that every human action should be chosen so that future generations too, including the seventh, will find a beautiful and livable earth.

The first Earth Day, celebrated in 1970, signaled that "environmental issues had begun to take an important place in the growing conscience of larger sectors of society." It was followed by a number of Earth Summits[A] which between 1972 and 2012 began to bring the global ecological crisis to the attention of nations and governments. Ecuador in 2008, and Bolivia in 2010, inspired by the ancient Indigenous cultures of the region, adopted the rights of Mother Earth (*Pachamama*) as the basis of a renewed federal constitution. In January 2017, Mexico amended its constitution to acknowledge "the wider spectrum of the rights of nature, comprising all its ecosystems and species as a collective entity, subject to its own rights." It also "recognizes animals as sentient beings, and therefore needing to be treated with dignity." (par. 3)[3]

An international conference on the rights of nature (Geneva 2016)[4] prepared for taking a Universal Declaration of the Rights of Mother Earth to UN level. Important concepts like ecocentric democracy (**ecodemocracy**[B]), and legal paths to stop ecocide were discussed.[5] In March 2017, the non-profit organization Nature's Rights (natures-rights.org) held a conference in Brussels to advance Earth jurisprudence in the EU. In the same month, the rivers Whanganui (New Zealand), Ganges, and Yamuna (both in India) were all granted legal personality and rights, as the first rivers to receive that status. Also in 2017, the IUCN[6] (the leading global authority on nature conservation) adopted rights of nature in its resolutions and work program. The United Nations' Harmony with Nature network has over 200 experts promoting Earth jurisprudence and nature's rights as a systemic solution to support the global transition.[7]

What can I do?

- "Be the voice for Mother Earth, Say yes to Rights of Nature."
  thepetitionsite.com/en-gb/1/yes-to-rights-of-nature/

Stopping ecocide

The degradation of the ecosphere continues, despite scores of international agreements such as codes of conduct, UN resolutions, treaties, protocols, etc. The problem is that none of these international agreements prohibits ecocide. Whatever ecological (or even ecocentric) program the UN will adopt will only be able to recommend, not enforce. What we need is a framework of actual ecocide *crime*, creating a legal duty of care that holds perpetrators against the living world to account in a criminal court of law.

The stage for this already exists. It is the International Criminal Court (ICC) in the Hague (Netherlands) and its governing document, The Rome Statute. Adopted in Rome in July 1998, it is one of the most powerful legal documents in the world. Crimes that already exist within the jurisdiction of the ICC are known collectively as Crimes Against Peace, they are:

1. the crime of genocide,
2. crimes against humanity,
3. war crimes,
4. the crime of aggression. But...
5. the crime of ecocide... is still missing! It was on the draft document in the mid-1990s but in a behind-closed-doors session disappeared last minute. (Imagine what a different world we would live in now, if it had remained.) The task now is to bring it back.

This will change our world entirely, and for the better.

Ecocide is a crime against the entire living natural world, not just against humans. Ecocide is a crime against the Earth. Currently there is a missing responsibility to protect. Ecocide law will reinstate our collective duty of care to protect the natural living world and all life. To cover the crime of ecocide under international law is a major step to protect the Earth.

The UK-based barrister, author and lead expert on ecocide law, Polly Higgins, founded an international ecocide criminal law response team of lawyers, forensic experts and former judges. She received a number of awards for her work advocating for a law of ecocide. She passed away on Easter Day 21 April 2019 after a rapid cancer but her team continues the work: stopecocide.earth.

As soon as the ecocide law will appear on the horizon, the effects will manifest quite swiftly, because bankers, insurance companies and investors will feel increasingly uneasy in supporting ecocidal projects like oil exploration, deforestation, ocean exploitation, etc. Among other things, the role of fossil fuels in ecosphere degradation will have to end pretty swiftly when ecocide law is established.

The future has long begun! In December 2018 began a revolutionary investigation when a Dutch minister and two Shell CEOs were "identified as principal suspects [in...] the potential crime of climate ecocide."[8] And in December 2019 two climate-vulnerable island states, Vanuatu and the Maldives, called for serious discussion of ecocide at the International Criminal Court. This sparked a wave of interest by international governments. And in January 2021, the European Parliament voted to urge "the EU and the Member States to promote the recognition of ecocide as an international crime under the Rome Statute of the International Criminal Court (ICC)". The official text emphasizes "that biodiversity and human rights are interlinked and interdependent, and recalls the human rights obligations of states to protect the biodiversity on which those rights depend".[9] The ball has finally started rolling...

What can I do?

- Visit Stop Ecocide at stopecocide.earth and support the cause by signing the international petition.
- If you can, also make donations to Stop Ecocide. All donations help to support the expansion of this global conversation.
- Become an Earth Protector. This is your visible support for the campaign. If you are an on-the-ground activist, it also offers a legal document to prove you are acting from your conscience to prevent harm, not to cause it.

Values

There is a lot of deep green potential in **religion** too. And also a need for transformation, as religions too got corrupted by *wétiko* hubris long ago. In history, the world faiths have sponsored wars or at least battles against the ecosphere, and still none of them challenges agribusiness, or human population overgrowth, which is inhumane and "a fundamental sin against the rest of creation," as Stan Rowe pointed out.[10] Particularly the monotheistic religions have been anthropocentric from the start. All the more surprising was Pope Francis' encyclical on the environment, released in June 2015. Titled *Laudato Si: On Care for Our Common Home,* it features outspoken statements about "pollution, waste and the throwaway culture", climate, water, and loss of biodiversity (par. 20-42). Still no mention of family planning, but it is a powerful and influential appeal for (almost) integral ecology, and a timely call to action.[11]

There is probably no wider and deeper exploration of the deep green potential of faiths than in Martin Palmer's work with the Association of Religions and Conservation (ARC) which began in 1995. The wide spectrum of ARC activities and initiatives includes:

- The Assisi Declarations, at which five leaders of the five major world religions—Buddhism, Christianity, Hinduism, Islam and Judaism—were invited to come and discuss how their faiths could help save the natural world. And since then, the ARC furthered:
- Islamic environmental education programs;
- the Muslim Seven Year Plan for a "long-term commitment to protect the living planet" (its programs including greening Medina, and setting up a green eco-labelling scheme);
- Green Pilgrimage Network handbooks for different faiths;
- assisting Daoism in "turning China's ecological crisis around";
- The Green Hindu Temples Guide;
- the Stewardship of God's Creation environmental toolkit for the Catholic University;
- women-led tree nurseries and entrepreneurship training, established by the Evangelical Lutheran Church of

Tanzania;
- a Green Award scheme for Islamic groups in the UK;
- assistance towards a Fatwa protecting threatened animals in Indonesia;
- assisting the foundation of EcoSikh, an organization which issued the first Sikh Statement on Climate Change, and organizes the annual Sikh Environment Day;
- "Judaism and Pollinators: Helping Pollinators Through Faith";
- roadmapping a "Low carbon future for Quakers".

Many more examples at ARC Downloads: http://www.arcworld.org/downloads.asp

From the ARC statement on the Global Alliance for Climate-Smart Agriculture (2014):

> Our faith partners believe we have a responsibility to protect the Living Planet because it was created by a Loving Creator and reveals the presence of the Divine, whatever we perceive the Divine to be. The natural world does not belong to a minority of rich or powerful individuals or corporations. Nor does it exist simply to serve humanity through providing "eco-system services" or other mechanistic roles. Instead, it is a manifestation of the Divine and is worth protecting for the sake of its own intrinsic value—including the fact that it is beautiful.[12]

Very different to "religion" is Indigenous spiritual life; it shouldn't be called "faith" or "religion" because it has nothing to do with believing. Widely referred to as **animism**, it is a way of life and set of practices. Animism (from *anima* meaning "soul") attributes "spirit" and "life" to all of nature. For Earth-conscious tribal peoples, animism is an active participation in the exchanges in nature. Stan Rowe emphasizes that "humanity needs to regain some form of animism for the preservation of its environment and itself,"[13] because animism, or Indigenous spirituality, is at its heart ecocentric. And Rowe quotes Fritjof Capra: "When the concept of the human spirit is understood as the mode of consciousness in which the individual feels a sense of belonging, of connectedness to the cosmos as a whole (and, my addition, "to the Earth in particular") it becomes clear that ecological awareness is spiritual in its deepest essence."[14]

Schooling

Western school systems are entirely outdated insofar they originate in the Industrial Age when the prime motive was to streamline children into a generic homogenized work force, obedient state subjects with few original thoughts of their own. To be herded into the kinds of jobs which are now disappearing because of cheap labor elsewhere and digitalization. Modern societies need the opposite: original individuals with bright ideas, creativity, innovation, head and heart. Buying computers for kindergartens is old thinking, and a crime against the children. They will master their IT skills soon enough, but first let them develop a firm base in reality, let them master their senses and motor skills, and root their being in harmony with planet Earth—what could be more important than to educate ourselves to become ecological citizens?[15] Facing the mess we brought upon us, what's the point of knowing details about historical kings and popes, or of Caesar's and Napoleon's conquests (and history's other celebrations of genocidal psychopaths who made it to the top).

All too often, schools can be among the most unimaginative places, and that's exactly the wrong way to trigger the potentials of children. "If you want your children to be intelligent," said Albert Einstein, "read them fairy tales. If you want them to be more intelligent, read them more fairy tales." Or, in the words of author Neil Gaiman: "We all have an obligation to daydream. We have an obligation to imagine. It is easy to pretend that nobody can change anything, that society is huge and the individual is less than nothing. But the truth is individuals make the future, and they do it by imagining that things can be different."[16]

How can we expect humanity to change if schools remain "hierarchical educational institutions where quantitative growth, exploitation, capitalist norms and environmental wastefulness are part of the group culture"? Prof. Alexander Lautensach from the School of Education of the University of Northern British Columbia, Canada, is pioneering the development of a **deep green curriculum**. Fully aware that climate disruption "will reduce agricultural productivity, biodiversity and public health, and rising sea levels will flood coastal lowland [...], driving unprecedented numbers of displaced people to find shelter in host communities with vastly different cultural traditions," Lautensach promotes a Transition curriculum. Instead of the conventional discourse about "security", "progress" and "growth" it attempts to counteract the anthropocentric conditioning, to pay explicit attention to ethics, and to encourage critical questioning. It will also extend a scientific worldview that embraces empathy and beauty in nature. The outmoded anthropocentric conception of human "security" is not supported by real conditions anymore (as its four pillars of socio-political, economic, environmental and public health structures are disintegrating). It needs replacement by *ecosphere security*. Since all life depends on the ecosphere, global ecosecurity is "the essential life-supporting 'space suit' for humanity." And with learning environments becoming increasingly multicultural, an ecocentric curriculum also offers equal bonding opportunities.[C] Adults also have to teach *by example*, of course.

What can I do?

- If you are a teacher, parent or carer, read the full article on the Transition curriculum in *The Ecological Citizen* vol 1, No 2: 171-8, at ecologicalcitizen.net

Circular economy

Just as the materials for products must flow in closed cycles (instead of endlessly producing garbage), money must also flow. The way giant companies in the dark anthropocentric age has been draining huge profits from regional communities dried up the prosperity of countries and their peoples. As Prof. Harald Lesch says: "Billionaires are dead ends for capitalism. Money is useless if it doesn't flow back."[17] There's a saying, "Once there was a man who was so poor, he had nothing but money." In the future rich people will no longer be celebrated. We will look at them with compassion because they have a lot of extra responsibility to carry.

Harmony

Listening is key to our ancient and ever-new relationship with nature. "Natural places have a harmony that embraces me when I set foot in them, a harmony of lives," says the environmental scientist Haydn Washington from the University of New South Wales in Sydney, Australia. "I feel this as both a scientist and a poet. Indeed, as a scientist, I cannot ignore what is so clear in such places. I do understand (as an ecologist) that that harmony is a dynamic equilibrium, where there is a state of flux... and yet the harmony endures. Indeed, if you listen, the harmony reaches out and teaches."[18]

"First and foremost, I believe it is harmony that we should be aiming for: it should be our mission, our goal, our vision and our path. We must seek an ethics of harmony, a true Earth ethic." Washington is by no means alone. The UN Harmony with Nature program (harmonywithnatureun.org) was adopted in December 2009. It states: "The Harmony with Nature initiative speaks to the need to move away from a human-centered worldview—or 'anthropocentrism'—and establish a non-anthropocentric, or Earth-centered, relationship with the planet. Under this new paradigm, Nature is recognized as an equal partner with humankind and is no longer treated as merely the source of raw materials to produce ever more commodities and feed the indefinite private accumulation of capital." This is a huge step for humanity, although the notion that we could possibly be "equal" with nature is yet another remnant of human hubris.

"Awareness of harmony goes hand in hand with a sense of wonder at life, the true love of the land," says Washington. "Many other concepts tie in with this notion of harmony, the most obvious being *respect* and *responsibility*. We must have the deepest respect and reverence for this evolved harmony of natural places. And we should feel a responsibility to maintain that harmony, and a duty to aid it and to celebrate its ongoing existence."

When they ask where I am, tell them:
Gone into leaf and stone and water
Gone into bone, and bird, and air.
– Roselle Angwin[19]

What can I do? (After Washington 2018)

- ***Be*** there with nature! Belong in the land.
- Take your children and friends to wild places so they can see the natural world as it really is, and bond with it.
- Take time to listen and ponder.
- Keep your imagination, creativity and artistic expression alive. They renew your sense of wonder.
- Cherish the imagination of your children, and let them

play in natural places (even small ones) with unstructured play.

Responsibility

Adults often call teenagers "spoilt brats" but in fact our entire culture is as decadent as it comes. People, movements, nations always demand their "rights" but avoid talking about responsibilities. "I have a right to eat my daily steak!", "I have a right to drive a gas-guzzling SUV!", "I have a right to do what I like!"—No, you don't! Not on an overpopulated planet anyway. Let's grow up!

An excellent perspective on the balance of rights and responsibilities is given by Sherri Mitchell/Weh'na Ha'mu'Kwasset who is a Native American attorney, author, teacher and activist from the Penobscot River in Maine. She says that the Indigenous word for law means "entering into kinship with another", and that all Indigenous rights are recognized as based upon the First Treaty which was "made with the creator when we first emerged into this world."[20]

"It is this treaty that gives us all of the rights that we claim under the laws of Man. In order to claim these rights we must honor the responsibilities that we accepted under that agreement. We're to live in harmony with the natural world and all living beings."

"This provides us with the foundational authority for all the rights that we stand on today. Under this agreement we have the right to live unencumbered on this land with full access to all of the sources of our survival. We don't view these things as 'resources'. They are the sources of our survival, they are not 'resources' to be exploited and commodified. Things such as water, food, shelter: we have the opportunity to enjoy these things as long as we uphold *our* responsibility to live in balanced harmony with the rest of creation."

"Without this balance we would destroy the very ground

that these rights are built upon. [...] The failure of industries to take responsibility for the destruction of our planet, and our complacency and complicity are quickly taking away our right to exist here on Mother Earth."

Sherri Mitchell emphasizes that the human rights are interdependent, tied to the rights of *all* of the other living beings that made their home here on Earth. And she explains that these principles also "translate into our individual lives and the work that we are doing collectively to protect both human and Earth-based rights." It is important never to forget that *our demand for rights must be balanced with a set of clear responsibilities, that we must be willing to do more than make a demand, we must be willing to actively work to create a world where that demand can be met:*

- "We can not claim a right to clean water without taking responsibility for actively ensuring that the contamination and over-use of the water is eliminated."
- "We can not claim a right to clean air without taking responsibility for the creation of healthy and sustainable energy sources and industrial practices."
- "We can not claim a right to a more equitable economy without taking responsibility for where and how we spend our money."

And her last point may also help to guide us regarding overpopulation and migration:

- "And we can not claim a right to life without taking responsibility for the lives that have already been created."

Footnotes

A. Earth Summits: Stockholm in 1972, Rio de Janeiro in 1992, Johannesburg in 2002, and Rio again in 2012.

B. Ecodemocracy aims to create decision-making systems that

respect the principles of human democracy, while explicitly recognizing the intrinsic value of non-human nature's rights. In votes, for example, human proxies can stand in for other species or ecosystems or their biotic or abiotic parts.[21]

C. "The ecocentric Transition necessitates a readiness to accept personal sacrifice and renouncement of privilege; this includes limitations of human rights that were universalized only recently in human history—moral territory that nobody gives up easily." (Alexander Lautensach 2018)[22]

Chapter 19

Finding Hope, Courage, and Strength

Research on social change shows that a major transformation does not require the entire population to acknowledge a problem. 2-3 percent of the people is enough, plus a strong grassroots movement and a few people in parliament. That is, for example, how the abolition movement ended slavery in the 1860s, or the suffragettes established voting rights for women in the early 20th century. Today, the numbers of Earth protectors are ever-growing, and so are the alliances between groups and movements. And we have the most powerful ally, as the naturalist Joe Gray reminds us: "We and the Earth are pushing in the same direction."[1]

Indeed, so many good projects and initiatives have been started, and more are popping up like mushrooms in autumn. There are so many solutions—for better ways, for more caring and ecologically sustainable ways of farming, economics, schooling, social and political life. There is a lot of light beginning to shine through the darkness.

But then your mood swings again: you see the powerful cannibalistic megamachine not giving an inch, and year after year business-as-usual grinding by while humanity is seriously running out of time. Hope based on facts has the tendency to wobble, to oscillate with our moods and the latest news that come to our attention. We need additional sources of strength, ones that are less dependent on changes in the physical world. We need to find strength and courage *within ourselves*.

Those who have faith are lucky. All of the world faiths have potent green seeds in their traditions (albeit sometimes very hidden) which can grow to wonderful plants of sustenance and Earth healing (see previous chapter). We may also remember

Sir George Trevelyan's words, "Don't underrate the prospect of divine intervention." However, faith and hope shall *never* keep us passive, because just standing by and watching the big destruction amounts to complicity.

What else is there? Reason can be helpful too. Before we ever give up, we should not forget that there have always been rapid leaps in the history of humanity. There were many war economies, and people knew how to adapt to them. The fact that an old regime is collapsing is usually only noticed by the rulers and the people when it actually happens. And the fact that no one can know in advance how the whole story will end is only natural. Maybe, humanity just *has to* learn the painful way. Then, in the words of ecocentric pioneer Patrick Curry, "the unfolding of the tale itself [is our] best hope for a happy 'ending'."[2]

"No matter how dark the future appears we must [...] refuse despair," says Curry,[3] and he's backed by Mary Robinson, the former president of Ireland: "Feeling 'This is too big for me, I give up' is no use to anybody. [With] despair, all the energy to do something goes out of the room."[4] Because despair is paralyzing and thus self-fulfilling. Patricia Espinosa, head of the UN Framework Convention on Climate Change, also warns of despair: "Falling into despair and hopelessness is a danger equal to complacency, none of which we can afford."[5]

"Despair," says Gandalf, "is for those who see the future beyond any doubt. We do not."[6] If you are a *Lord of the Rings* fan (and who isn't), your recipe against despair is clear: in the face of overwhelming adversity and no reasonable ground for hope, you need the courage, the authenticity, and the resilience of Frodo, and in the end what will save the day is the humbleness and devotion of Sam Gamgee.

Another pitfall is misanthropy: why not hate the human species, the agent of all this destruction? Because we cannot! If we cherish Earth and all her children, how could we exclude our own species "from the care and concern we try to extend to all the others?" Furthermore, deeming our race uniquely horrible is yet another disguised anthropocentrism: we are so special, one way or the other; or both! And also, says Curry, "it is far from true that all humans are equally destructive. So damning them all is not only unfair, it drives away those whom we need on our—that is, the Earth's—side."[7]

We have no guarantees, and no constant to hold on to. We cannot know if civilization, or humanity, or even the ecosphere will be saved from the last fatal blow. We have to try our best, because: "who knows? But it has got to be worth a try, because if we don't then the answer will certainly turn out to be 'no'. Even if we do try, there are no guarantees; but then there is a chance."[8]

At the end of the day, it all amounts to finding our "appreciation of life itself, at once natural and spiritual, as the ultimate value."[9] We don't need any ideology, and not even hope. We only do *the right thing*. Those who stand up for the "right thing" don't do it because they think it will work. They do it because it is the right thing to do.

During the occupation of Wounded Knee in 1973, the leader of the American Indian Movement (AIM), Russell Means, was asked what kept him going against all odds, and he answered, "We are not concerned about winning or losing. We are not concerned about the overwhelming odds against us. We are here taking a stand because it is the right thing to do and the right place to do it."[10] And Captain Watson says: "We must be prepared to take a stand in the present to make it a better world for tomorrow. I am always optimistic because I believe that the answer to a seemingly impossible problem is to find an impossible solution and I believe that imagination and courage driven by passion is the path to finding impossible solutions."[11]

And nothing is more inspiring and encouraging than love. There is so much worth living for. "Even in a world of dying," says writer Jonathan Franzen, "new loves continue to be born."[12]

"Love is a mystery. So in one sense it does not matter whether the Earth is alive or not. Our love for her is something we give. And in return she gives us her love," says First Nation activist and writer Jack D. Forbes.[13] As it is, Earth sustains us all, day by day. And like our foremothers and forefathers on all continents, we can see Gaia as a mighty being who generates wonder and awe, respect and gratitude, joy and love inside our soul.

Afterword

First Nations activist Sherri Mitchell says: "It is time for us all to step into our role as ancestors of the future and dream the next seven generations into being. In doing so, we must recognize that giving them life is not enough. We must also work to provide them with a world that is capable of sustaining their lives. This is the work of our time, the work of our lives. To succeed in this work, we must reconnect with the thread of life. We must learn the spiritual language spoken by our ancestors and renew the relationship that they long held with the rest of creation. Together, we must [...] write a new story based on cooperation and conscious co-creation of a more humane and sacred way of being."[1]

If my daughter was in serious trouble I would want to be there for her. And if need be, lay down my life so she could live. That is in every way preferable to not having been there when she needed me most. Where there is love and responsibility, there is no space for fear. Or despair. Or denial. There is only caring, and doing what needs to be done. Same with Earth: she has been taken hostage by bands of dangerous lunatics, and I want to be there for her in her darkest hour, with all of you, my sisters and brothers, and all our relatives. If I could choose again the time of my incarnation in this mortal coil, the peaceful fields of Arcadia in ancient Greece or a sensual, joyful Indigenous life on a Southern island long ago would tempt my wary soul. But I would decide for this moment, the dark, troubled early 21st century with its contaminated elements, interrupted cycles, mauled life support systems, and disemboweled biosphere. When human hubris and supremacy have unleashed hell on Earth, then She needs all her lovers most.

That's what it's about: We arrive HERE, finally, fully, with all our heart. If we step into ourselves like this, and allow the flames of our passionate love for all-that-lives to leap high, let ourselves be transformed like the phoenix, then we can be an active part of birthing a future that is livable for *all beings*. A healthy and happy planet is beginning right now!

Gaia, mother of all,
the foundation, the oldest one,
I shall sing to Earth.

She feeds everything in the world.

Whoever you are,
whether you walk upon her sacred ground,
or move through the paths of the sea,
you who fly,
it is she who nourishes you from her treasure store.

from *Hymn to Gaia* by Hesiod (fl. ca. 750 BCE)

Glossary

Anthropocene

"Age of Man". The name of a new geological epoch that is also a warning cry. Mass extinction, climate collapse and planetary contamination by heavy metals, radioactivity and plastic particles have begun to leave their traces in geological layers.

Anthropocentrism

The greatest problem of humankind. The view that humans are the epicenter of all being and becoming and that all creatures and systems have to serve us. Usually accompanied by the denial that other living beings have inherent beauty and dignity and the right to exist regardless of the extent to which they serve humanity. Opposed to this is *ecocentrism*.

Biodiversity

The diversity of biological species and their complex interactions. It also includes the genetic diversity within species and the diversity of ecosystems. Compare *species diversity*

Biome

The biological community (flora and fauna) that lives and evolves together in a certain natural environment based on similarities in regional climate and terrain, geographic location, and other characteristics. The four main terrestrial biomes are tundra, forest, grassland and desert, each having sub-divisions. The major freshwater biomes include ponds and lakes, streams and rivers, and wetlands. Marine biomes include coral reefs, estuaries, and the ocean biome—the largest of all of the Earth's biomes.

Biocapacity

Earth's biological power to regenerate and reproduce plant matter, renew particular resources (e.g. freshwater reservoirs) as well as assimilate waste. This primary productivity of nature is the source for all animal (and hence human) life, and is increasingly threatened by pollution, soil erosion, freshwater loss, and climate disruption. Compare *Carrying capacity*

Carbon Budget

In the context of climate policy an upper limit of total CO_2 emissions associated with remaining below—with a certain probability—a specific global average temperature. Since the beginning of industrialization, humanity has already emitted about 1,900Gt (metric gigatons) of CO_2; to have a two-thirds chance of keeping global heating below 1.5 degrees Celsius, the remaining permissible amount would be about 420Gt (IPCC Special Report 2018). At annual emissions of 38Gt, the entire CO_2 budget will be used up before 2030, while politicians continue to *promise* a halving of CO_2 emissions by 2030 (and climate neutrality by 2050) but fail to act accordingly. Other greenhouse gases are calculated separately by the IPCC. The unknown quantities of feedback loops and tipping points, however, are difficult to take into account.

Carbon dioxide equivalent (CO_2e)

A way to place emissions of various greenhouse gases on a common footing by accounting for their effect on climate. It states for a given amount of a greenhouse gas the amount of CO_2 that would have the same effect on global heating (when measured over a specified time period). CO_2e is often calculated assuming a 100-year global warming potential (GWP100) of CO_2. See *Global warming potential*

Carrying capacity
The ability of a regional ecosystem or the planet's ecosphere to indefinitely support the *optimum population* of a given species in balance, without degradation of the whole ecosystem. In the case of the human species it also depends on the levels of consumption/pollution and technology. See also *Ecological Footprint*; *Overshoot*

Climate
Long-term average of weather (30 years or more) in a larger area. The surface of the Earth is covered by a pattern of different climates, which, however, influence each other. The climatic processes mainly regulate the water management of the ecosphere. The driving energy source is the sun. Anthropogenic global heating has a strong impact on all climates. In the discussion, climate is succinctly spoken of in the singular, as if there were a single world climate.

Degrowth
A planned contraction of human economic activity towards a sustainable, equitable steady state within the means of Earth's ecosphere. The underlying insight is that there cannot be *infinite* (economic) growth on a *finite* planet.

Doughnut economics
One of the new concepts for an economy that is ecologically sustainable. Developed by economist Kate Raworth, the Doughnut Model is a visual framework for "sustainable development" which acknowledges the reality of planetary boundaries as well as social boundaries, i.e., avoids both ecological overshoot and social deficits. Compare *War on nature*

Ecocentrism
The view that everything that occurs in nature—organisms,

species, places, landscapes, ecosystems—has inherent dignity and thus the right to exist and flourish according to its own nature. "Its own nature" also includes its balanced role in the ecosystem of which it is part. In the community of life on Earth, Homo sapiens is but one species among many, without special rights. Opposed to this is *anthropocentrism*.

Ecocide

A criminal human activity that causes the serious harm or destruction of an ecosystem or of any species therein (including humans). Ecocide is the missing crime in the Rome Statute of the International Criminal Court; its inclusion has the power to significantly change humanity's current course of self-destruction.

Ecological Footprint

A way of measuring humanity's demand on the ecosphere on which it is entirely dependent. The Footprint of a human population is the biologically productive land and water area required to produce the resources it consumes and absorb the waste it generates. Because today people consume resources and ecological services from all over the world, their Footprint is the sum of these areas, wherever they are on the planet. The US American way of life for 8 billion people would require four Earths. See also *Carrying capacity*; *Overshoot*

Ecology

The science of interactions among living organisms and their natural environment which together form ecosystems. Ecology does not look at the individual components of nature (organisms, elements, matter, energy) but at how these parts interact in a system. It is essential to understand that organisms cannot survive without their ecosystem, because they are "open systems" too.

Ecosphere
The planetary envelope in which all biological life exists. The terms ecosphere and biosphere are interchangeable, with biosphere emphasizing the living, and ecosphere emphasizing its multiple relationships. For our better understanding, we subdivide into biomes, ecosystems, communities, etc., but in reality it is a single, multi-layered interwoven entity that makes all of our lives possible.

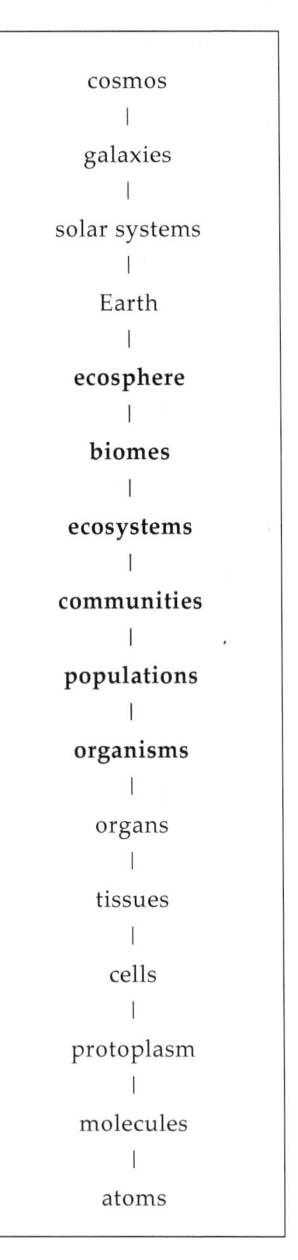

Levels of organization
The bold levels are studied by ecology. (after Erle Ellis)

Ecosystem
A multitude of organisms interacting with each other and with their environment in ways that energy and matter are exchanged. An ecosystem is a level above that of the ecological community, and a level below, or equal to, biomes. Ecosystems evolve to be sustainable systems, but they need a continuous flow of high-quality energy to maintain their structure and function. This energy comes from the sun. All ecosystems are therefore "open systems". See *Biome*

Geoengineering
Suggestions for global-scale technological interventions with the planet, proposing to offset the

effects of climate breakdown by reducing atmospheric carbon. All of the proposed measures (e.g. huge mirrors in Earth's orbit to reflect the sunlight; filtering atmospheric CO_2 and burying compressed carbon in the soil) would be impossibly expensive and, as agreed by the majority of scientists, resemble dangerous global experiments with too many unknowns. The only reason geoengineering keeps coming up in the discussion is that the fossil fuel industry uses its promises of CO_2 reduction as a delaying tactic to prevent *real* climate action.

Global warming potential

To enable comparison of different greenhouse gases, their global warming potential is calculated over a time span, usually a century (GWP100) or twenty years (GWP20). In the case of methane, for example, GWP100 makes it appear more benign than GWP20, because methane is about 34 times more climate-active than CO_2 over a century, but 86 times more climate-active during the first two decades in the atmosphere. Since about 2015, climate scientists have been urging policy makers to switch to GWP20. See *Carbon dioxide equivalent*

Greenwashing

A term used to charactize product advertisements and labels that promise more environmental benefit than they actually have. Also applies to politicians selling business-as-usual projects as "green solutions". Greenwashing cases span from being ridiculous to being unethical to covering up health hazards to downright facilitation of ecocide. Ultimately, "cosmetic environmentalism" such as Greenwashing is a symptom of governments and industries still resisting the growing pressure to take real responsibility for the ecosphere and its web of life.

Overshoot

Exploiting the ecosphere beyond the Earth's regenerative

capacity and filling natural waste sinks to overflowing, i.e. dumping (often toxic) waste beyond nature's processing capacity. By 1999, humanity had reached 120 percent of the carrying capacity of planet Earth, i.e. the human enterprise was 20 percent in ecological overshoot. The planet is being depleted, and the human growth economy is running ever faster towards global ecological collapse. See also *Carrying capacity*; *Ecological Footprint*

Regenerative agriculture

Tilling agricultural land, and also overgrazing by livestock, causes soil degradation and releases carbon from the soil. The difference between tilling and not tilling is about 1 ton of pure carbon per acre and year. Over millennia of agriculture, humanity has created the so-called "soil carbon debt", which denotes the amount of carbon missing from the depleted soils of the world. Regenerative farming uses techniques that sequester carbon, rebuild soil, increase its biological life as well as its mineral content, protect the purity of groundwater, and reduce erosion.

Rewilding

An emerging global movement aiming to return land from human control to nature, thereby strengthening communities and ecosystems and rebuilding biodiversity.

Species diversity

The diversity of biological species in an ecosystem or in the entire ecosphere of the earth. Compare *biodiversity*

Sustainability

The ability to continue in perpetuity. In ecology, to avoid overharvesting, overfishing and any other form of Overshoot, both in taking from nature's treasures as well as in discarding wastes. Every thriving species eventually encounters limits to growth set by time, space, and energy, and humanity has reached its absolute limits around the last turn of the century. *Sustainability is the absolute opposite of and incompatible with endless economic growth*. Eventually, sustainability has to entail ending the war on nature, and the inevitability of the associated path of self-destruction. See *War on nature*

War on nature

One of the underlying principle paradigms of our culture, based on the fundamental misinterpretation of the Biblical "subdue the Earth", amplified by the Royal Academy of Sciences which in the late 17th century aimed to make nature subservient to mankind's purposes in order to achieve "the Empire ('Dominion') of Man over Nature" and over "the inferior creatures of God". This fuelled the development of unprecedented greed and the history of ecocides. The myth that everything in nature is war penetrates modern medicine as much as industrial agriculture, both wielding arsenals of antibiotics ("against life") and mass destruction (pesticides etc.), both having created multi-resistant super-bugs. The "war on corona" is the latest manifestation of choosing a scapegoat ("the evil virus") instead of addressing the underlying causes (in this case, the exploitation of animals for food, and also tremendous levels of air pollution annually killing millions of people—and wild animals—via infections of the air passages. See *Sustainability*

Weather

Local events in the atmosphere: solar radiation, cloud structure, temperature, atmospheric pressure, humidity, precipitation, air movement.

For more, visit the Iowa State University's "Encyclopedia of Earth": https://editors.eol.org/eoearth

Source References

The same list of source references but with full URLs can be found on the author's website at https://www.healthyplanet.one/references-en

Prelims

1. UN News 2020. COVID-19 pandemic, an "unprecedented wake-up call" for all inhabitants of Mother Earth. 22 April.
2. United Nations 2020. UN International Mother Earth Day 22 April Message.
3. UN Environment 2019. *Global Environment Outlook—GEO-6: Healthy Planet, Healthy People*. Cambridge University Press.
4. Jack D. Forbes 2008 [1992, 1979]. *Columbus and Other Cannibals*. New York: Seven Stories Press. 181-2.
5. Hippocrates. *On Airs, Waters, and Places*.

Part I: Living Earth

1. Alfred Russel Wallace 1876. *The Geographical Distribution of Animals, vol. 2*. 553.

Chapter 1: Foundations

1. Eileen Crist and H. Bruce Rinker 2010. One Grand Organic Whole. In Eileen Crist and H. Bruce Rinker, eds., *Gaia in Turmoil: Climate Change, Biodepletion, and Earth Ethics in an Age of Crisis*. Cambridge MA/London: MIT Press. 3-20.
2. ibid.
3. James Lovelock 2000a. *Gaia: A New Look at Life on Earth*. Oxford University Press. 19-20, 30.
4. Dyna Rochmyaningsih 2012. Include trees in climate modelling, say scientists: Climate models should include the effects of trees on the local climate, say agroforestry experts. *The Guardian Environment Network*, 16 January.

5. Tim Flannery 2012. *Here on Earth: A Twin Biography of the Planet and the Human Race*. London: Penguin. 46.
6. James Lovelock 2000c, 111, referring to DW Schwartzman and T. Volk (1989), Biotic enhancement of weathering and the habitability of Earth. *Nature* 340, 457-460.
7. Tim Flannery 2012, op. cit. 45.
8. AM Makarieva and VG Gorshkov 2006. Biotic pump of atmospheric moisture as driver of the hydrological cycle on land. *Hydrology and Earth System Sciences* 3, 2621-2673. See also: Peter Bunyard 2010. The Real Importance of the Amazon Rain Forest. ISIS Report, 15/03/2010.
9. Elisabet Sahtouris 1996. The Gaia controversy: a case for the Earth as an evolving organism. 324-338, in Bunyard, ed., 1996. *Gaia in Action: Science of the Living Earth*. Edinburgh: Floris. 331.
10. Stephan Harding, Lynn Margulis 1996. Water Gaia: 3.5 thousand million years of wetness on planet Earth. In Eileen Crist and H. Bruce Rinker, eds., 2010. op. cit. 41-60.
11. All quotes in Elisabet Sahtouris 1996. The Gaia controversy: a case for the Earth as an evolving organism. 324-338, in Peter Bunyard, ed., 1996. *Gaia in Action: Science of the Living Earth*. Edinburgh: Floris. 332.
12. Gregory J. Hinkle 1996. Marine salinity: Gaian phenomenon? In Bunyard 1996. op. cit. 99-104.
13. ibid.
14. Tim Flannery 2012, op. cit. 46.

Chapter 2: Origins

1. Stephan Harding 2009. *Animate Earth: Science, Intuition and Gaia*, 2nd ed. Cambridge (UK): Green Books. 109.
2. e.g. *Archaeoglobus fulgidus*; compare Wikipedia: Sulfate-reducing microorganism.
3. Still today, cyanobacteria "contribute significantly to global ecology and the oxygen cycle" (Wikipedia: Cyanobacteria).

4. James Lovelock 2000b, op. cit. 110; James Lovelock 2000c, op. cit. 79-81.
5. Tim Flannery 2012, op. cit. 48-9.
6. Wikipedia: Late Paleozoic icehouse.
7. Stephan Harding 2009, 210.

Chapter 3: The Elements and Cycles

1. Stephan Harding 2009. *Animate Earth: Science, Intuition and Gaia*, 2nd ed. Cambridge (UK): Green Books. 96.
2. Stephan Harding 2009, op. cit. 99.
3. James Lovelock 2000c, op. cit. 115.
4. Jobin Jacob and K. Suthindhiran 2016. Magnetotactic bacteria and magnetosomes – Scope and challenges. *Materials Science and Engineering: C*. 68.
5. Joseph L. Kirschvink, Atsuko Kobayashi-Kirschvink, Barbara J. Woodford 1992. Magnetite Biomineralization in the Human Brain. *Proceeding of the National Academy of Sciences of the United States of America* volume 89, issue 16, pp. 7683-7687, 08/1992.
6. James Lovelock 2000b, op. cit. 73, 121.
7. AM Makarieva and VG Gorshkov 2006, op. cit.
8. James Lovelock 2000c, op. cit. 119.
9. Stephan Harding 2009, op. cit. 178.
10. S. Dyhrman, J. Ammerman, B. Van Mooy 2007. Microbes and the marine phosphorus cycle. *Oceanography* 20:110-116.
11. James Lovelock 2000b, op. cit. 133-41; James Lovelock 2000c, op. cit. 122-5.
    Compare Wikipedia: Dimethyl sulfide.
    Also Wikipedia: CLAW hypothesis.
12. Stephan Harding 2009. *Animate Earth: Science, Intuition and Gaia*, 2nd ed. Cambridge (UK): Green Books. 114-119, 140-144.
    See also: Wikipedia: Coccolithophore.
    And: Wikipedia: Emiliania huxleyi.

Chapter 4: Communities and Networks

1. Jaime A. Teixeira da Silva & Judit Dobránszki 2016. Magnetic fields: how is plant growth and development impacted? *Protoplasma* volume 253, 231-248.
2. Valeriy Zaporozhan, Andriy Ponomarenko 2010. Mechanisms of Geomagnetic Field Influence on Gene Expression Using Influenza as a Model System: Basics of Physical Epidemiology. *Int J Environ Res Public Health,* 2010 March; 7(3): 938-965.
3. Joseph L. Kirschvink, James L. Gould 1981. Biogenic magnetite as a basis for magnetic field detection in animals. *Biosystems* volume 13, issue 3, 1981, 181-201.
4. Chao-Hung Liang, Cheng-Long Chuang, et al. 2016. Magnetic Sensing through the Abdomen of the Honey bee. *Scientific Reports* volume 6, 23657.
5. V. Lambinet, ME Hayden, et al. 2017. Linking magnetite in the abdomen of honey bees to a magnetoreceptive function. *Proc. R. Soc.* B284: 20162873.
6. Ietse Stokroos, et al. 2001. Keystone-like crystals in cells of hornet combs. *Nature* volume 411, 654.
7. Christine Maira Hein, Svenja Engels, et al. 2011. Robins have a magnetic compass in both eyes. *Nature* volume 471, E1.
8. J. Roger Brothers, Kenneth J. Lohmann 2018. Evidence that Magnetic Navigation and Geomagnetic Imprinting Shape Spatial Genetic Variation in Sea Turtles. *Current Biology* volume 28, issue 8, 1325-1329. April 23.
9. Yinon M. Bar-On, Rob Phillips, Ron Milo 2018. The biomass distribution on Earth. *PNAS,* June 19, 2018, 115 (25) 6506-6511.
10. Angelicque E. White 2009. New insights into bacterial acquisition of phosphorus in the surface ocean. *Proc Natl Acad Sci USA,* 2009 December 15; 106(50): 21013-21014.
11. Stephan Harding 2009. *Animate Earth: Science, Intuition and*

Gaia, 2nd ed. Cambridge (UK): Green Books. 163.

12. Arthur Prindle, Jintao Liu, et al. 2015. Ion channels enable electrical communication in bacterial communities. *Nature* volume 527, 59-63.
13. A. Widom, J. Swain, et al. 2012. Electromagnetic Signals from Bacterial DNA. Cornell University.
14. Stephan Harding 2009. *Animate Earth: Science, Intuition and Gaia*, 2nd ed. Cambridge (UK): Green Books. 165.
15. Stephan Harding 2009, op. cit. 166.
16. Harding 1996, op. cit. 193.
17. ibid.
18. "Slime Mold Physarum Finds the Shortest Path in a Maze". Excerpt of the award-winning German docu *Als wären sie nicht von dieser Welt – Der unmögliche Lebenswandel der Schleimpilze* (2002) on YouTube.
19. *Slime Mould outperforming Humans*. Excerpt of the BBC docu (2011) on YouTube.
20. Alan Rayner, in Harding 2009, op. cit. 195.
    Also: Paul Stamets 2005. *Mycelium Running: How mushrooms can help save the world*. New York: Random House. 1-7.
21. Harding 1996, op. cit. 205f.
22. André Scheffel, Manuela Gruska, et al. 2006. An acidic protein aligns magnetosomes along a filamentous structure in magnetotactic bacteria. *Nature* volume 440, 110-114.

Chapter 5: Feedback Systems

1. ME Marsh 2003. Regulation of $CaCO_3$ formation in coccolithophores. *Comparative Biochemistry and Physiology B* 136 (4): 743-754.
2. L. Beaufort, et al. 2011. Sensitivity of coccolithophores to carbonate chemistry and ocean acidification. *Nature* 476 (7358): 80-3.
3. Aldo Leopold 1949. *A Sand County Almanac: And Sketches Here and There*. Oxford University Press.

4. GrrlScientist 2014. How Wolves Change Rivers – video. theguardian.com, 3 March.
Recommended: "How Wolves Change Rivers – video" (4:33). YouTube. https://www.youtube.com/watch?v=ysa5OBhXz-Q Note that the narrator in this video refers to the prey animals as deer, which is correct insofar wapiti (Cervus canadensis) belongs to the deer family (Cervidae), but confusing because North Americans call it "elk". Which in turn would be confusing to British English speakers because the animal they call elk (Alces alces) is called moose in America.
5. Holger Dambeck 2018. Wie viele Wölfe verträgt das Land? spiegel.de, 01.06.
6. Karl A. Mayer, M. Tim Tinker, et al. 2019. Surrogate rearing a keystone species to enhance population and ecosystem restoration. Cambridge University Press, *Fauna & Flora International*.
7. Isabelle Groc 2020. Furry engineers: sea otters in California's estuaries surprise scientists. theguardian.com, 14 August.

Chapter 6: Diversity, Complexity, and Abundance

1. Eileen Crist 2010. Intimations of Gaia. In Eileen Crist and H. Bruce Rinker, eds., 2010. *Gaia in Turmoil: Climate Change, Biodepletion, and Earth Ethics in an Age of Crisis*. Cambridge MA/London: MIT Press. 315.
2. James Lovelock 2004. Reflections on Gaia. In S. Schneider, J. Miller, E. Crist, and P. Boston, eds., *Scientists Debate Gaia: The Next Century*. Cambridge (MA): The MIT Press. 1-5.
3. Eileen Crist 2010, op. cit. 315.
4. Eileen Crist and H. Bruce Rinker 2010. One Grand Organic Whole. In Eileen Crist and H. Bruce Rinker, eds., 2010, op. cit. 3-20.
5. James Lovelock 2000b, op. cit. 23.
6. Eileen Crist 2010. Intimations of Gaia. In Eileen Crist and

H. Bruce Rinker, eds., 2010, op. cit. 328f.

Part II: Global Disruption

1. António Guterres 2020. A Time to Save the Sick and Rescue the Planet. nytimes.com, April 28.
2. Naomi Klein 2020. How big tech plans to profit from the pandemic. theguardian.com, 13 May.
3. William E. Rees 2020. Ecological economics for humanity's plague phase. *Ecological Economics* volume 169, March 2020, 106519.
4. Kate Raworth 2017. *Doughnut Economics: Seven Ways to Think Like a 21st-Century Economist*. Random House Business.
   Watch her Ted Talk on YouTube (17 min): Why it's time for 'Doughnut Economics' | Kate Raworth | TEDxAthens.
   Or a short animated introduction here: 1. Change the Goal - 1/7 Doughnut Economics. YouTube.
   A more scientific approach: University of Leeds: A Good Life For All Within Planetary Boundaries.
5. Quoted in George Monbiot 2017. Finally, a breakthrough alternative to growth economics – the doughnut. theguardian.com, 12 April.
6. Nicola Davison 2019. The Anthropocene epoch: have we entered a new phase of planetary history? theguardian.com, 30 May.
7. Higgins, P., Short, D., & South, N. Protecting the planet: a proposal for a law of ecocide. *Crime Law Soc Change* (2013) Vol 59, 257.
8. Boris Worm, et al. 2013. Global catches, exploitation rates, and rebuilding options for sharks. *Marine Policy* 40(1):194-204, July 2013. DOI: 10.1016/j.marpol.2012.12.034.
9. Wikipedia: Ecocide.

Chapter 7: The Sixth Mass Extinction

1. IPBES Global Assessment Report 2019. B4, 4.

2. Jonathan Watts 2018. Destruction of nature as dangerous as climate change, scientists warn. theguardian.com, 23 March.
Also: Jonathan Watts 2018. Habitat loss threatens all our futures, world leaders warned. theguardian.com, 17 November.
3. Damian Carrington 2018. What is biodiversity and why does it matter to us? theguardian.com, 12 March.
4. Subhankar Banerjee 2018. Biological Annihilation: a Planet in Loss Mode. *CounterPunch*, December 13.
5. Jonathan Watts 2018. Global warming should be called global heating, says key scientist. theguardian.com, 13 December.
6. Eileen Crist 2010. Intimations of Gaia. In Eileen Crist and H. Bruce Rinker, eds. *Gaia in Turmoil: Climate Change, Biodepletion, and Earth Ethics in an Age of Crisis*. Cambridge MA/London: MIT Press. 315-333.
See also: Gerardo Ceballos, Paul R. Ehrlich, Anthony D. Barnosky, Andrés García, Robert M. Pringle and Todd M. Palmer 2015. Accelerated modern human-induced species losses: Entering the sixth mass extinction. *Science Advances*, 19 June, Vol. 1, no. 5, e1400253.
7. Crist 2010, op. cit.
8. Wade Davis 2018. On Ecological Amnesia. *The Tyee*, 8 November.
9. Eileen Crist 2010. Intimations of Gaia. In Eileen Crist and H. Bruce Rinker, eds., 2010, op. cit. 315-333.
10. WWF Living Planet Report 2018.
11. Daniel G. Boyce, Marlon R. Lewis, Boris Worm 2010. Global phytoplankton decline over the past century. *Nature* 466 (7306): 591-6. Bibcode:2010Natur.466..591B.
12. World Fish Migration Foundation 2020. The Living Planet Index (LPI) for migratory freshwater fish.
13. WWF 2020. WWF Living Planet Report 2020: Bending the

curve of biodiversity loss. Gland, Switzerland.

14. MJ Samways 2019. Addressing global insect meltdown. *The Ecological Citizen* 3 (Suppl A): 23-6.
15. WWF 2018. Living Planet Report 2018, op. cit.
16. IPBES Global Assessment Report 2019. A3, 2-3.
17. Prof. Dave Goulson quoted in: Damian Carrington 2018. What is biodiversity and why does it matter to us? theguardian.com, 12 March.
18. Agence France-Presse 2018. Giant African baobab trees die suddenly after thousands of years. theguardian.com, 11 June.
19. Chronic oak dieback: https://www.forestresearch.gov.uk/tools-and-resources/pest-and-disease-resources/chronic-oak-dieback/

    Acute oak decline: https://www.forestresearch.gov.uk/tools-and-resources/pest-and-disease-resources/acute-oak-decline/
20. Michigan State University 2013. Managing Dothistroma and brown needle blight on pines.
21. Tom Horton 2011. Revival of the American Chestnut. americanforests.org, December 1st.
22. Subhankar Banerjee 2018. Biological Annihilation: a Planet in Loss Mode. *CounterPunch*, December 13. Referring to the Forest Ecology and Management report 2010.
23. Patrick Greenfield 2020. UN draft plan sets 2030 target to avert Earth's sixth mass extinction. theguardian.com, 13 January.
24. EO Wilson (interview) 2010. Edward O. Wilson: The loss of biodiversity is a tragedy. *Media Services*, 09.02.2010.
25. WWF 2018. Living Planet Report 2018: Aiming higher.
26. Daniel G. Boyce, et al. 2014. Estimating global chlorophyll changes over the past century. *Progress in Oceanography* 122: 163-73.
27. Damian Carrington 2017. Warning of 'ecological

Armageddon' after dramatic plunge in insect numbers. theguardian.com, 18 October.
28. Alejandra Borunda 2018. Italy's Olive Trees Are Dying. Can They Be Saved? nationalgeographic.com, August 10. Also: Alison Abbott 2018. Italy's olive crisis intensifies as deadly tree disease spreads. nature.com, 13 November.
29. Ash dieback (Hymenoscyphus fraxineus). https://www.forestresearch.gov.uk/tools-and-resources/pest-and-disease-resources/ash-dieback-hymenoscyphus-fraxineus/
30. Bleeding Canker of Horse Chestnut (Pseudomonas syringae pv aesculi). https://www.forestresearch.gov.uk/tools-and-resources/pest-and-disease-resources/bleeding-canker-of-horse-chestnut/

Chapter 8: Habitat Destruction

1. Stephan Harding 2010. Gaia and Biodiversity. In Eileen Crist and H. Bruce Rinker 2010, 109.
2. IPBES Global Assessment Report 2019. 17.
   And: International Rivers. https://www.internationalrivers.org/questions-and-answers-about-large-dams
3. Patrick Barkham 2018. Europe faces 'biodiversity oblivion' after collapse in French birds, experts warn. theguardian.com, 21 March.
4. Damian Carrington 2018. What is biodiversity and why does it matter to us? theguardian.com, 12 March.
5. Leyland Cecco 2018. Canada's salmon hold the key to saving its killer whales. theguardian.com, 12 November.
6. Dom Phillips 2018. Brazil records worst annual deforestation for a decade. theguardian.com, 24 November.
7. Fiona Harvey 2020. Amazon near tipping point of switching from rainforest to savannah – study. theguardian.com, 5 October.
   Arie Staal, Ingo Fetzer, et al. 2020. Hysteresis of tropical forests in the 21st century. *Nature Communications* volume

11, 4978.
8. Hannah Ellis-Petersen 2020. India plans to fell ancient forest to create 40 new coalfields. theguardian.com, 8 August.
9. Martua T. Sirait 2009. Indigenous Peoples and Oil Palm Plantation Expansion in West Kalimantan, Indonesia. Amsterdam University Law Faculty.
10. Abrahm Lustgarten 2018. Palm Oil Was Supposed to Help Save the Planet. Instead It Unleashed a Catastrophe. nytimes.com, November 20.
11. Martua T. Sirait 2009, op. cit.
12. Arthur Neslen 2016. Protected forests in Europe felled to meet EU renewable targets – report. theguardian.com, 24 November.
13. Subhankar Banerjee 2018. Biological Annihilation: a Planet in Loss Mode. *CounterPunch*, December 13, 2018.
14. Tom Horton 2011. Revival of the American Chestnut. americanforests.org, December 1st.
15. Subhankar Banerjee 2018, op. cit. Referring to a study by the Los Alamos National Laboratory: April Reese 2018. Bird Population Plummets in Piñon Forests Pummeled by Climate Change. *Audubon*, August 14.
16. Robin McKie 2019. 'To save our fish, we must first find ways to unblock UK's rivers,' say scientists. theguardian.com, 1 September.
17. Joshua Jones, et al. 2019. A comprehensive assessment of stream fragmentation in Great Britain. *Science of The Total Environment* volume 673, 10 July 2019, 756-762.
18. Jonathan Watts 2019. Human society under urgent threat from loss of Earth's natural life. theguardian.com, Mon 6 May.
19. Stephan Harding 2010, op. cit. 107-124. Referring to Bert C. Klein 1989. Effects of Forest Fragmentation on Dung and Carrion Beetle Communities in Central Amazonia. *Ecology*

volume 70, No. 6 (December 1989), Ecological Society of America.

20. Convention on Biological Diversity 2020. Global Biodiversity Outlook 5. Montreal.
21. Jonathan Watts 2018. Habitat loss threatens all our futures, world leaders warned. theguardian.com, 17 November.
22. Quoted in: Jonathan Watts 2018. China urged to lead way in efforts to save life on Earth. theguardian.com, 29 November.
23. Quoted in: Jonathan Watts 2018. Stop biodiversity loss or we could face our own extinction, warns UN. theguardian.com, 6 November.

    Also: UN Decade on Biodiversity. Press Release, 29 November 2018.
24. Leyland Cecco 2018. Canada's salmon hold the key to saving its killer whales. theguardian.com, 12 November.
25. Recommended: Edward O. Wilson 2016. *Half-Earth: Our Planet's Fight for Life*. New York: Norton.
26. EO Wilson quoted in: Robin McKie 2018. Should we give up half of the Earth to wildlife? theguardian.com, 18 February.
27. The Half-Earth Project: https://www.half-earthproject.org/
28. E. Dinerstein, et al. 2020. A "Global Safety Net" to reverse biodiversity loss and stabilize Earth's climate. *Science Advances*, 04 September 2020: Vol. 6, no. 36, eabb2824.

    A good interactive presentation: Niko Kommenda 2020. Planetary 'safety net' could halt wildlife loss and slow climate breakdown. theguardian.com, 29 September.
29. Rewilding Charter Working Group (2020) Global Charter for Rewilding the Earth. *The Ecological Citizen* 4(Suppl A): 6-21.
30. Greenpeace Global Oceans: https://www.greenpeace.org.uk/what-we-do/oceans/
31. "Rang-Tan in my bedroom" on YouTube.

32. Stephan Harding 2010, op. cit. 109.
33. WWF 2018. Living Planet Report 2018: Aiming higher.
34. Leyland Cecco 2018, op. cit.
35. Dom Phillips 2018, op. cit.
36. Jonathan Watts 2019. Amazon deforestation 'at highest level in a decade'. theguardian.com, 18 November 2019.
37. Sean Sloan, et al. 2018. Hidden challenges for conservation and development along the Trans-Papuan economic corridor. *Environmental Science and Policy* 92 (2019) 98-106.

Chapter 9: Invasive Species

1. Lidia Chitimia-Dobler, et al. 2016. First detection of *Hyalomma rufipes* in Germany. *ScienceDirect, Ticks and Tick-borne Diseases* 7, Issue 6, October, 1135-1138.
   Also: WHO 2013. Crimean-Congo haemorrhagic fever. 31 January.
2. Conal Urquhart 2015. Red squirrel finds pine marten a fearsome ally in its fight for survival. theguardian.com, 22 February.
3. S. Dubois, N. Fenwick, EA Ryan, et al. 2017. International consensus principles for ethical wildlife control. *Conservation Biology* 31: 753-60. In EC Parke and JC Russell 2018. Ethical responsibilities in invasion biology. *The Ecological Citizen* 2: 17-19.
4. D. Fraser 2012. A 'practical ethic' for animals. *Journal of Agricultural and Environmental Ethics* 25: 721-46. In EC Parke and JC Russell 2018. Ethical responsibilities in invasion biology. *The Ecological Citizen* 2: 17-19.
5. David Derbyshire 2018. The terrifying phenomenon that is pushing species towards extinction. theguardian.com, 25 February.
6. Subhankar Banerjee 2018. Biological Annihilation: a Planet in Loss Mode. *CounterPunch*, December 13.
   Also: Lynda V. Mapes 2014. Starfish are 'just melting':

Disease killing 80 percent of them. seattletimes.com, June 30, 2014.
Also: Lynda V. Mapes 2016. Scientists now link massive starfish die-off, warming ocean. seattletimes.com, February 21.
7. Banerjee 2018, op. cit.
8. Australian Associated Press 2018. Sydney suffers through hottest day since 1939 as temperatures reach 47.3C in Penrith. theguardian.com, 7 January.
9. Deutschlandfunk 2018. Bachforellensterben – Forscher finden Ursache des rätselhaften Phänomens. DLF, Umwelt & Verbraucher, 29.11.
10. Simon Romero 2020. New Mexico Mystery: Why Are So Many Birds Dropping Dead? nytimes.com, September 15.
11. Jeremy Plester 2017. All hell breaks loose as the tundra thaws. theguardian.com, 20 July.
Compare: Wikipedia: Permafrost > Microbes.
12. Jim Robbins 2012. The Ecology of Disease. nytimes.com, July 14.
13. Marcia L. Kalish, et al. 2005. Central African Hunters Exposed to Simian Immunodeficiency Virus. *Emerg Infect Dis*, 2005 December; 11(12): 1928-1930.
14. Jim Robbins 2012, op. cit.
15. Jim Robbins 2012, op. cit.
16. Peter Beaumont 2019. What's really behind the spread of Lyme disease? Clue: it's not the Pentagon. theguardian.com, 19 July.
17. Lidia Chitimia-Dobler, et al. 2016, op. cit.
18. Quoted in Robbins 2012, op. cit.
19. Mattha Busby 2020. 'Live animals are the largest source of infection': dangers of the export trade. theguardian.com, 21 January.
20. Rob Wallace (interview) 2020. Where did coronavirus come from, and where will it take us? *Uneven Earth*, March 12.

21. ibid.
22. The IUCN List of Threatened Species. https://www.iucnredlist.org/
    And: Global Register of Introduced and Invasive Species. http://www.griis.org/about.php
    Jonathan Watts 2018. New global registry of invasive species is 'milestone' in protecting biodiversity. theguardian.com, 23 January.
23. James C. Scott 2018. *Against the Grain: A Deep History of the Earliest States*. Yale University Press.
24. Colin J. Carlson 2020. From PREDICT to prevention, one pandemic later. *The Lancet*, Comment, Volume 1, ISSUE 1, e6-e7, May 01.

Chapter 10: Pollution

1. Damian Carrington 2019. Microplastics 'significantly contaminating the air', scientists warn. theguardian.com, 14 August.
   Also: Cheryl Katz 2019. Tiny pieces of plastic found in Arctic snow. nationalgeographic.com, August 14.
2. JL Pauly, et al. 1998. Inhaled cellulosic and plastic fibers found in human lung tissue. American Association for Cancer Research, Volume 7, Issue 5, 419-428.
3. Kieran D. Cox, et al. 2019. Human Consumption of Microplastics. American Chemical Society, *Environ. Sci. Technol.*, 2019, 53, 12, 7068-7074, June 5.
4. ibid.
5. Dunzhu Li, et al. 2020. Microplastic release from the degradation of polypropylene feeding bottles during infant formula preparation. *Nature Food*, 19 October.
   Damian Carrington 2020. Microplastics revealed in the placentas of unborn babies. theguardian.com, 22 December.
6. Roland Geyer, Jenna R. Jambeck, and Kara Lavender Law 2017. Production, use, and fate of all plastics ever made.

Science Advances, 2017 July; 3(7): e1700782. Published online 2017 July 19.

7. Adam Vaughan 2016. Biodegradable plastic 'false solution' for ocean waste problem. theguardian.com, 23 May.
8. Fiona Harvey 2020. Plastic waste entering oceans expected to triple in 20 years. theguardian.com, 23 July.
9. Sandra Laville 2018. Tyres and synthetic clothes 'big cause of microplastic pollution'. theguardian.com, 22 November.
10. N. Evangeliou, et al. 2020. Atmospheric transport is a major pathway of microplastics to remote regions. *Nature Communications* volume 11, 3381.
11. Ian Sample 2018. 'Sad surprise': Amazon fish contaminated by plastic particles. theguardian.com, 16 November.
12. United Nations Environment Programme 2018. How to banish the ghosts of dead fishing gear from our seas. www.unenvironment.org, accessed in December.
13. Laura Paddison 2016. Single clothes wash may release 700,000 microplastic fibres, study finds. theguardian.com, 27 September.
14. Damian Carrington 2017. Plastic fibres found in tap water around the world, study reveals. theguardian.com, 6 September.
15. Andrew JR Watts, et al. 2015. Ingestion of Plastic Microfibers by the Crab Carcinus maenas and Its Effect on Food Consumption and Energy Balance. *Environ. Sci. Technol.* 49 (24), 14597-14604.
16. Leyland Cecco 2018. Canada's salmon hold the key to saving its killer whales. theguardian.com, 12 November.
17. John Vidal 2016. Microplastics should be banned in cosmetics to save oceans, MPs say. theguardian.com, 24 August.
18. Fiona Harvey 2016. Microplastics killing fish before they reach reproductive age, study finds. theguardian.com, 2 June.

19. Hannah Summers 2018. Great Pacific garbage patch $20m cleanup fails to collect plastic. the guardian.com, 20 December.
20. Matthew Taylor 2017. Plastics found in stomachs of deepest sea creatures. theguardian.com, 15 November.
21. Damian Carrington 2018. Plastic pollution discovered at deepest point of ocean. theguardian.com, 20 December.
22. Damian Carrington 2018. Microplastics can spread via flying insects, research shows. theguardian.com, 19 September.
23. Damian Carrington 2017. Plastic fibres found in tap water around the world, study reveals. theguardian.com, 6 September.
24. Alicia Mateos-Cárdenas, et al. 2020. Rapid fragmentation of microplastics by the freshwater amphipod *Gammarus duebeni* (Lillj.). *Scientific Reports* volume 10, 12799.
25. Government Office for Science. Foresight Future of the Sea: A Report from the Government Chief Scientific Advisor.
26. See video clip "WWF report warns annihilation of wildlife threatens civilisation – video". 30 October 2018.
27. Pressestelle Medizinische Universität und Wien Pressestelle Umweltbundesamt 2018. Erstmals Mikroplastik im Menschen nachgewiesen.
28. Fiona Harvey 2016, op. cit.
29. Fiona Harvey 2018. Whale and shark species at increasing risk from microplastic pollution – study. theguardian.com, 5 February.
30. Damian Carrington 2018. Orca 'apocalypse': half of killer whales doomed to die from pollution. theguardian.com, 27 September.
    Also: Monika Seynsche 2016. PCB hat katastrophale Folgen für Meeressäuger. Deutschlandfunk.de, 19. Juni.
31. Leyland Cecco 2018. Canada's salmon hold the key to saving its killer whales. theguardian.com, 12 November.

32. BUND 2001. Hormonaktive Substanzen im Wasser – Gefahr für Gewässer und Mensch.
33. Oliver Milman 2018a. 'Great Pacific garbage patch' sprawling with far more debris than thought. theguardian.com, 22 March.
34. Wikipedia: Phase-out of lightweight plastic bags.
35. Jennifer Rankin 2019. European parliament votes to ban single-use plastics. theguardian.com, 27 March.
36. Dream McClinton 2019. Florida bill would ban plastic straw bans until 2024. theguardian.com, 6 March.
37. BBC News. Plastic straws: Government confirms ban in England. bbc.co.uk, 22 May 2019.
38. France 24. La France a commandé près de 2 milliards de masques en Chine. 04/04/2020.
39. Ashifa Kassam 2020. 'More masks than jellyfish': coronavirus waste ends up in ocean. theguardian.com, 8 June.
40. WHO 2020. Advice on the use of masks in the context of COVID-19: Interim guidance. 5 June.
41. Sandra Laville 2020. New British standard for biodegradable plastic introduced. theguardian.com, 1 October.
42. EPA superfund: https://www.epa.gov/superfund
Also: Chris D'Angelo 2020. EPA's Superfund program, a Trump priority, is in shambles. grist.org, January 11.
43. Damian Carrington 2018. Orca 'apocalypse': half of killer whales doomed to die from pollution. theguardian.com, 27 September.
44. Sandra Laville 2018. Tyres and synthetic clothes 'big cause of microplastic pollution'. theguardian.com, 22 November.
45. Louise Edge, oceans campaigner at Greenpeace UK, quoted in Laura Paddison 2016. Single clothes wash may release 700,000 microplastic fibres, study finds. theguardian.com, 27 September.
46. YouTube: "The Man Clearing 9,000 Tons of Trash From

Mumbai's Beaches."
47. United Nations Environment Programme. How to banish the ghosts of dead fishing gear from our seas. www.unenvironment.org, accessed in December 2018.
48. United Nations Environment Programme. Tackling marine debris. www.unenvironment.org, accessed in December 2018.
49. Fiona Harvey 2016. Scientists call for better plastics design to protect marine life. theguardian.com, 14 July.
50. jme/dpa 2018. Der größte Mikroplastik-Verursacher sind Autoreifen. spiegel.de, 04.09.
51. Rebecca Smithers 2020. Device to curb microplastic emissions wins James Dyson award. theguardian.com, 17 September.
52. Wikipedia: Cradle-to-Cradle.
53. Steinar Brandslet 2020. Bioplastics no safer than other plastics. norwegianscitechnews.com, 10.20.
54. According to the conservative EU Member of Parliament, Karl-Heinz-Florenz, in: Paul Vorreiter: EU-Parlament diskutiert über Plastikverbot. DLF, 23.10.2018.
55. YouTube: "Trashed trailer".
56. Wikipedia: Nanotechnology.
57. Fabienne Schwab quoted in: Kurt F. de Swaaf 2011. Nano-Müll könnte Leben im Wasser stören. spiegel.de, 13.11. See: Fabienne Schwab, et al. 2011. Are Carbon Nanotube Effects on Green Algae Caused by Shading and Agglomeration? *Environ. Sci. Technol.*, 2011, 45, 14, 6136-44. And: Fabienne Schwab, et al. 2013. Diuron Sobbed to Carbon Nanotubes Exhibits Enhanced Toxicity to *Chlorella vulgaris*. *Environ. Sci. Technol.*, 2013, 47, 13, 7012-7019.
58. YouTube: "More Than Honey – Official Trailer".
59. Ian Whyte 2018. Life's catastrophe: An angry editorial. *The Ecological Citizen* 2: 5-10. Referring to: Human Rights Council 2017. Report of the Special Rapporteur on the

Right to Food. UN General Assembly, New York, NY, USA.

60. Damian Carrington 2017. UN experts denounce 'myth' pesticides are necessary to feed the world. theguardian.com, 7 March.
61. UN 2019. Global Environmental Outlook – GEO-6: Healthy Planet, Healthy People.
62. Dipali Singh, Ladislav Nedbal, Oliver Ebenhöh 2018. Modelling phosphorus uptake in microalgae. *Biochemical Society Transactions*, April 17, 2018, 46 (2) 483-490.
    UN 2021. Launch of the Second World Ocean Assessment (WOA II).
63. https://www.umweltbundesamt.de/faqs-zu-nitrat-im-grund-trinkwasser#textpart-3
64. Julia Merlot 2018. Darum geht es in der Gülleklage. spiegel.de, 21.06.
65. Margaret McCasland, et al. 2012. Nitrate: Health Effects in Drinking Water. Cornell University, Pesticide Safety Education Program (PSEP).
66. Sandra Ratzow 2018. Neuseeland: Verseuchtes Trinkwasser. daserste.de, 23.04.
67. Department of Agriculture, Fisheries and Food 2010. Food Harvest 2020: A vision for Irish agri-food and fisheries.
68. Ella McSweeney 2020. 'We've crossed a threshold': has industrial farming contributed to Ireland's water crisis? theguardian.com, 28 September.
69. Chris Arsenault 2014. Only 60 years of farming left if top soil degradation continues. *Scientific American*, December 5.
70. Volkert Engelsman quoted in: Chris Arsenault 2014, op. cit.
71. Recommended reading: Damian Carrington 2018. Global food system is broken, say world's science academies. theguardian.com, 28 November.
    Compare: Philip Case 2014. Only 100 harvests left in UK farm soils, scientists warn. *Farmers Weekly*, 21 October 2014.

72. BUND 2001. Hormonaktive Substanzen im Wasser – Gefahr für Gewässer und Mensch.
73. Fred Pearce 2002. Dung to Death. *New Scientist*, 20 April 2002, 20, referring to Stephan Mueller of the Swiss Federal Institute of Aquatic Science and Technology (Eawag).
74. Bündnis 90/Die Grünen 2019. Eklatante Mängel in der Schweinehaltung. Referring to: Albert Sundrum. Tierschutzmängel in der Schweinehaltung – Erläuterungen zum aktuellen Stand.
75. Fiona Harvey 2019. Superbug hotspots emerging in farms across globe – study. theguardian.com, 19 September 2019.
76. Fred Pearce 2002, op. cit.
77. Gregory Robinson 2019. Antibiotic resistance rising among dolphins, study reveals. theguardian.com, 15 September.
78. Ernst Ulrich von Weizsaecker, Anders Wijkman, et al. 2018. *Come On! Capitalism, Short-termism, Population and the Destruction of the Planet. A Report to the Club of Rome.* New York: Springer. ch. 1.6.1.
79. SynBioWatch 2016. 160 Global Groups Call for Moratorium on New Genetic Extinction Technology at UN Convention.
80. SynBioWatch 2016. The Gene Drive Files: Disclosed Emails Reveal Military as top funder of dangerous new genetic manipulation technique.
81. Maria Varenikova 2020. Chernobyl Wildfires Reignite, Stirring Up Radiation. nytimes.com, April 11.
82. Bundesamt für Strahlenschutz 2018. Radioaktive Belastung von Pilzen und Wildbret. 22.10.
83. Mark Diesendorf 2015. Accidents, waste and weapons: nuclear power isn't worth the risks. *The Conversation*, May 18.
84. Nicole Feldman 2018. The steep costs of nuclear waste in the U.S. Stanford University. Stanford Earth, July 03.
85. Duncan Clark 2012. How much do we spend on nuclear waste? theguardian.com, 16 November.

86. W. Richardson 2013. *Marine Mammals and Noise*. San Diego, CA: Academic Press.
87. Agence France-Presse 2017. Baby whales 'whisper' to mothers to avoid predators, study finds. theguardian.com, 26 April.
88. Christin T. Murphy, et al. 2017. Seal Whiskers Vibrate Over Broad Frequencies During Hydrodynamic Tracking. *Scientific Reports* 7(1), December.
89. Greenpeace 2003. Sonar tötet Wale. https://www.greenpeace.de/themen/meere/sonar-toetet-wale
90. PD Jepson, et al. 2013. What Caused the UK's Largest Common Dolphin (Delphinus delphis) Mass Stranding Event? *PLoS ONE* 8(4): e60953. https://doi.org/10.1371/journal.pone.0060953.
91. YouTube: "Wegen künstlichem Sonar: US-Paranoia vor russischen U-Booten gefährdet Wale". RT news, October 28, 2015.
92. Pippa Howard, et al. 2020. An assessment of the risks and impacts of seabed mining on marine ecosystems. *Fauna & Flora International*, March 2020.
93. Karen McVeigh 2020. David Attenborough calls for ban on 'devastating' deep sea mining. theguardian.com, 12 March.
94. Spain creates Mediterranean Sea reserve for whale migration. *Associated Press*, June 29 2018.
95. ibid.
96. EC Wietzikoski Lovato, PAG Velasquez, et al. 2018. High frequency equipment promotes antibacterial effects dependent on intensity and exposure time. *Clin Cosmet Investig Dermatol*, 2018; 11: 131-135.
97. SV Larionov, DV Krivenko, AV Avdeenko 2011. Effect of electromagnetic radiation of the extremely high frequency millimeter range on thermophilic cultures of bacteria of lactic acid products. *Russian Agricultural Sciences* volume 37, 434-435.

98. Diana Soghomonyan, et al. 2016. Millimeter waves or extremely high frequency electromagnetic fields in the environment: what are their effects on bacteria? *Appl Microbiol Biotechnol*, June 2016; 100 (11): 4761-71.
99. Zoë L. Hutchison, et al. 2020. Anthropogenic electromagnetic fields (EMF) influence the behaviour of bottom-dwelling marine species. *Scientific Reports* volume 10, article 4219.
100. Malka N. Halgamuge, et al. 2015. Reduced Growth of Soybean Seedlings After Exposure to Weak Microwave Radiation From GSM 900 Mobile Phone and Base Station. *Bioelectromagnetics* 36 (2), January.
101. Turs Selga, Maija Selga 1996. Response of *Pinus sylvestris L.* needles to electromagnetic fields. Cytological and ultrastructural aspects. *Science of The Total Environment* volume 180, issue 1, 2 February 1996, 65-73.
102. Katie Haggerty 2010. Adverse Influence of Radio Frequency Background on Trembling Aspen Seedlings: Preliminary Observations. *International Journal of Forestry Research* volume 2010, 836278.
103. Susan McGrath 2014. Cracking Mystery Reveals How Electronics Affect Bird Migration. nationalgeographic.com, May 7.
104. Kirill Kavokin, et al. 2014. Magnetic orientation of garden warblers (Sylvia borin) under 1.4 MHz radiofrequency magnetic field. *Journal of the Royal Society*, 06 August 2014.
105. Alfonso Balmori 2005. Possible Effects of Electromagnetic Fields from Phone Masts on a Population of White Stork (*Ciconia ciconia*). *Electromagnetic Biology and Medicine* volume 24, Issue 2.
106. Neelima R. Kumar, Sonika Sangwan, Pooja Badotra 2011. Exposure to cell phone radiations produces biochemical changes in worker honey bees. *Toxicol Int.*, 2011 Jan-Jun; 18(1): 70-72.
107. Christine Merlin, Robert J. Gegear, Steven M. Reppert

2009. Antennal circadian clocks coordinate sun compass orientation in migratory monarch butterflies. *Science* 325 (5948): 1700-1704.

108. Arno Thielens, Duncan Bell, et al. 2018. Exposure of Insects to Radio-Frequency Electromagnetic Fields from 2 to 120 GHz. *Scientific Reports* volume 8, 3924.
109. Recommended: Mark Hertsgaard and Mark Dowie 2018. The inconvenient truth about cancer and mobile phones. theguardian.com, 14 July.
110. Leif G. Salford, et al. 2003. Nerve Cell Damage in Mammalian Brain after Exposure to Microwaves from GSM Mobile Phones. *Environmental Health Perspectives* volume 111, number 7, June 2003, 881-883.
111. World Health Organization 2011. IARC classifies radio frequency electromagnetic fields as possibly carcinogenic to humans. Press release N° 208, 31 May.
112. SK Mishra, R. Chowdhary, S. Kumari, SB Rao 2017. Effect of Cell Phone Radiations on Orofacial Structures: A Systematic Review. *J Clin Diagn Res*, 2017 May;11(5):ZE01-ZE05.
113. Mark Hertsgaard and Mark Dowie 2018, op. cit.
114. Yinon M. Bar-On, Rob Phillips, Ron Milo 2018. The biomass distribution on Earth. *PNAS*, June 19, 2018, 115 (25) 6506-6511.
115. Jean Louis Guénet 2005. The mouse genome. Cold Spring Harbor Laboratory Press. *Genome Res*, 2005, 15: 1729-1740.
116. TED talk on YouTube: "How bacteria 'talk' – Bonnie Bassler".
117. joe/dpa 2018. Erstmals Mikroplastik in menschlichen Stuhlproben nachgewiesen. spiegel.de, 23.10.
118. All quotes Dr. Frank Borower (Hamburg, Germany): personal communication, January 2019.
119. Alice Ross 2018. UK household plastics found in illegal dumps in Malaysia. *Greenpeace Unearthed*, 21.10.

120. Ian Sample 2018. 'Sad surprise': Amazon fish contaminated by plastic particles. theguardian.com, 16 November.
121. Hannah Summers 2018. Great Pacific garbage patch $20m cleanup fails to collect plastic. theguardian.com, 20 December.
122. Jens-Peter Marquardt 2018. Zigaretten schädigen Lunge und Umwelt. DLF, Umwelt & Verbraucher, 25.10.
123. Jürgen Roth 2017. "Wir brauchen eine mittlere Katastrophe", Interview mit Peter Berthold. fr.de, 3. Dez.
124. Patrick Barkham 2018. Europe faces 'biodiversity oblivion' after collapse in French birds, experts warn. theguardian.com, 21 March.
125. USGS 2016. US Geological Survey Minerals Yearbook – Phosphate Rock.
126. DVGW 2015. Anthropogene Spurenstoffe in Gewässern. DVGW-Information Wasser Nr. 54, April.
127. Danny Hakim 2016. Doubts About the Promised Bounty of Genetically Modified Crops. nytimes.com, October 29.
128. Press Association 2017. Seals are deafened in noisy shipping lanes, say scientists. theguardian.com, 2 May.
Also: Jim Robbins 2018. Oceans Are Getting Louder, Posing Potential Threats to Marine Life. *New York Times* online, January 22, 2019.
129. Damian Carrington 2013. Whales flee from military sonar leading to mass strandings, research shows. theguardian.com, 3 July.
130. Leyland Cecco 2018. Canada: locals angry after navy holds live fire exercises in orca habitat. theguardian.com, 22 November.
131. J. Dreport 2018. 5G experimenten in Den Haag: honderden vogels vallen massaal dood uit de bomen. 2 November.
132. Anne Katharina Zschocke 2014. *Darmbakterien als Schlüssel zur Gesundheit – Neueste Erkenntnisse aus der Mikrobiom-Forschung*. Munich: Knaur. 22f.

133. Anne Katharina Zschocke 2014, op. cit. 41-72.
134. Recommended: The award-winning documentary *Microbirth: Revealing the microscopic events during childbirth.* 60 min. http://microbirth.com/
135. Anne Katharina Zschocke 2014, op. cit. 76–85.

Chapter 11: Population

1. Recommended: Philip Cafaro and Eileen Crist 2012. *Life on the Brink: Environmentalists Confront Overpopulation.* University of Georgia Press.
2. United Nations 2015. World Population Prospects: Key findings and advance tables. The 2015 Revision. New York.
3. Most famously Paul Ehrlich in *The Population Bomb* (Ballantine Books, New York 1968). But in hindsight his warnings are being revalued, for example:
   – Paul A. Murtaugh 2015. Paul Ehrlich's Population Bomb Argument Was Right. nytimes.com, June 8.
   – Damian Carrington 2018. Paul Ehrlich: 'Collapse of civilisation is a near certainty within decades'. theguardian.com, 22 March.
4. David Von Drehle 2019. Don't fall for the doomsday predictions. washingtonpost.com, January 8.
5. Tom Butler and Musimbi Kanyoro 2015. Burning down the house. Resilience.org, July 11.
6. Eileen Crist, Camilo Mora, Robert Engelman 2017. The interaction of human population, food production, and biodiversity protection. *Science* 356, 260-264 (2017) 21 April.
7. Crist, et al. 2017, op. cit.
8. ibid.
9. ibid.
10. United Nations 2015. World Population Prospects: Key findings and advance tables. The 2015 Revision. New York.
11. Karin Kuhlemann 2018. 'Any size population will do?': The fallacy of aiming for stabilization of human numbers. *The*

Ecological Citizen 1: 181-9.

12. Charlotte McDonald 2015. How many Earths do we need? BBC News, 16 June.
13. Patrick Curry 2011. *Ecological Ethics: An introduction*. Cambridge, UK: Polity Press. 253-257.
    David Pimentel, Michele Whitecraft, et al. 2010. Will Limited Land, Water, and Energy Control Human Population Numbers in the Future? *Human Ecology* volume 38, 599–611.
14. Rory Carroll 2018: Mary Robinson on climate change: "Feeling 'This is too big for me' is no use to anybody". theguardian.com, 12 October.
15. Glen Barry 2015. Europe's Refugee Crisis: Mass Migration is Biosphere Collapse. *EcoInternet*, September 13.
16. Brian Kahn 2016. Syria's drought 'has likely been its worst in 900 years'. theguardian.com, 2 March.
    Also: Colin P. Kelley, et al. 2015. Climate change in the Fertile Crescent and implications of the recent Syrian drought. *PNAS*, March 17, 112 (11) 3241-3246.
    Also: Jan Selby, et al. 2017. Climate change and the Syrian civil war revisited. *Political Geography* 60, September 2017, 232-244.
17. Robin McKie 2018. Should we give up half of the Earth to wildlife? theguardian.com, 18 February.
18. Colin Hines 2018. Immigration and population: The interlinked ecological crisis that dares not speak its name. *The Ecological Citizen* 2: 51-5.
19. Hannes Weber 2018. Germany's Cities and Their Environmental Footprint Are Growing Again. *NewSecurityBeat*, December 10.
20. German Federal Government 2016. German Sustainable Development Strategy: New Version 2016. 154.
21. German Federal Government 2016, op. cit. 154.
22. Wikipedia: List of countries and dependencies by

population density.
23. Colin Hines 2017. Progressive protectionism – the Green case for controlling our borders. *The Ecologist*, 13th January.
24. Wade Davis 2018. On Ecological Amnesia. *The Tyee*, 8 November.
25. Tom Butler and Musimbi Kanyoro 2015, op. cit.
26. John Bongaarts and Brian C. O'Neill 2018. Global warming policy: Is population left out in the cold? *Science* 361, 17 August, Issue 6403.
27. Tom Butler and Musimbi Kanyoro 2015, op. cit.
28. Quoted in Jürgen Stryjak 2018. Innenminister beraten – Wie gefährlich ist Syrien für heimkehrende Flüchtlinge? DLF, 28.11.
29. Oliver Milman, Emily Holden, and David Agren 2018. The unseen driver behind the migrant caravan: climate change. theguardian.com, 30 October.
30. Ahmad Salkida 2012. Africa's vanishing Lake Chad. UN *Africa Renewal*, April.
31. Mélanie Gouby 2020. Chad halts lake's world heritage status request over oil exploration. theguardian.com, 24 September.
32. Colin Hines 2018, op. cit.
33. Jon Clifton 2013. More Than 100 Million Worldwide Dream of a Life in the U.S. Gallup, March 21.

Chapter 12: Overconsumption

1. Eileen Crist 2010. Intimations of Gaia. In Eileen Crist and H. Bruce Rinker, eds. 2010. *Gaia in Turmoil: Climate Change, Biodepletion, and Earth Ethics in an Age of Crisis*. Cambridge MA/London: MIT Press. 315-333.
2. Jessica Aldred 2016. Agriculture and overuse greater threats to wildlife than climate change – study. theguardian.com, 10 August.
3. Damian Carrington 2020. World's consumption of materials

hits record 100bn tonnes a year. theguardian.com, 22 January.

4. Anne Herrberg 2018. Nicht nur in Peru: Schmutzige Goldförderung zerstört Ökosysteme. DLF, Umwelt & Verbraucher, 9 January.
5. Wikipedia: cocaine > economics.
6. Vernadsky quoted in: Stephan Harding and Lynn Margulis 2010. Water Gaia: 3.5 thousand million years of wetness on planet Earth. In Eileen Crist and H. Bruce Rinker 2010, op. cit. 41-59.
7. Barbara Harwood 2010. Gaia's freshwater: an oncoming crisis. In Eileen Crist and H. Bruce Rinker 2010, op. cit. 151-163.
8. UN 2019. Große Ungleichheiten beim Zugang zu Wasser. Unesco-Pressemitteilung, 13. März.
   Also: Karin Kuhlemann 2018. 'Any size population will do?': The fallacy of aiming for stabilization of human numbers. *The Ecological Citizen* 1: 181-9. Referring to FAO, 2009; 2012.
9. UN 2019, op. cit.
   Compare: FAO 2009. How to Feed the World in 2050: Issue brief. FAO, Rome, Italy.
   Also: FAO 2012. Coping with Water Scarcity: An action framework for agriculture and food security (FAO Water Reports). FAO, Rome, Italy.
10. Sondhya Gupta 2019. England's running out of water – and privatisation is to blame. theguardian.com, 21 March.
11. Brian Merchant 2009. How many gallons of water does it take to make… treehugger.com, June 24.
12. Fermín Koop 2017. Time running out for WTO to act on fishing subsidies. chinadialogue, 07.12.
    See also: Suzanne Goldenberg 2014. Fuel subsidies 'drive fishing industry's plunder of the high seas'. theguardian.com, 24 June.

13. Robin McKie 2017. North Atlantic's greatest survivors are hunted once more. theguardian.com, 25 November.
14. IUCN 2020. Almost a third of lemurs and North Atlantic Right Whale now Critically Endangered - IUCN Red List. 09 July.
15. Robin McKie 2018a. The oceans' last chance: 'It has taken years of negotiations to set this up'. theguardian.com, 5 August.
16. Mike Hoffman, of the Zoological Society of London, quoted in Robin McKie 2018b. Should we give up half of the Earth to wildlife? theguardian.com, 18 February.
17. George Monbiot 2015. We're treating soil like dirt. It's a fatal mistake, as our lives depend on it. theguardian.com, 25 March.
18. Richard Allison 2014. UK soil crisis hitting crop yields, warns expert. *Farmers Weekly*, 10 February.
19. Jürgen Roth 2017. "Wir brauchen eine mittlere Katastrophe", Interview mit Peter Berthold. fr.de, 3. Dez.
20. J. Sanderman, et al. 2017. Soil carbon debt of 12,000 years of human land use. *Proceedings of the National Academy of Sciences*.
21. IAP 2018. Opportunities for future research and innovation on food and nutrition security and agriculture: The InterAcademy Partnership's global perspective. InterAcademy Partnership, November. 45.
    Also: Chris Arsenault 2014. Only 60 years of farming left if top soil degradation continues. *Scientific American*, December 5.
22. Chris Arsenault 2014, op. cit.
23. Department for Environment, Food and Rural Affairs (DEFRA) 2015. Cross compliance in England: soil protection standards.
24. IUCN. Land degradation and climate change.
25. Paige L. Stanley, et al. 2018. Impacts of soil carbon

sequestration on life cycle greenhouse gas emissions in Midwestern USA beef finishing systems. *Agricultural Systems* 162, May 2018, 249-258.

26. Tim Flannery 2011, op. cit. 261-5.
27. Community Supported Agriculture (CSA): https://communitysupportedagriculture.org.uk/what-is-csa/
28. Bibi van der Zee 2018. What is the true cost of eating meat? theguardian.com, 7 May.
29. Dave Merrill and Lauren Leatherby 2018. Here's How America Uses Its Land. bloomberg.com, July 31.
30. Damian Carrington 2018. Humans just 0.01% of all life but have destroyed 83% of wild mammals – study. theguardian.com, 21 May.
31. Kelly Anthis 2019. Global Farmed & Factory Farmed Animals Estimates. Sentience Institute, February 21.
32. J. Poore, T. Nemecek 2018. Reducing food's environmental impacts through producers and consumers. *Science,* 01 June 2018: Vol. 360, Issue 6392, 987-992.
33. The Food and Land Use Coalition 2019. The Global Consultation Report, September.
34. Rob Wallace (interview) 2020. Where did coronavirus come from, and where will it take us? *Uneven Earth,* March 12.
35. Véronique Bouvard, et al. 2015. Carcinogenicity of consumption of red and processed meat. *The Lancet,* Oncology, volume 16, issue 16, 1599-1600, December 01.
36. Johns Hopkins Medicine 2018. Beef Jerky and Other Processed Meats Associated with Manic Episodes. *Neuroscience News,* July 21.
37. Marco Springmann, et al. 2020. Health-motivated taxes on red and processed meat: A modelling study on optimal tax levels and associated health impacts. *PLoS ONE* 13(11): e0204139.
38. Philipp Lichterbeck 2017. Mit Pfeil und Bogen gegen die Auslöschung. Cicero, 3. Mai.

39. French documentary *Tomorrow: Take concrete steps to a sustainable future*. A film by Cyril Dion and Mélanie Laurent. 2015.
40. Zac Goldsmith, UK MEP, in *The Economics of Happiness*. Films for Action. Min. 21.21.
41. *The Economics of Happiness*. Films for Action. Min. 22.00. And: Local Futures. Just how insane is trade these days? Factsheet.
42. Helen Harwatt 2018. Including animal to plant protein shifts in climate change mitigation policy: a proposed three-step strategy. *Climate Policy* volume 19, issue 5, tandfonline.com, 26 November.
43. Bibi van der Zee 2018, op. cit.
44. The EAT-Lancet Commission on Food, Planet, Health.
45. Damian Carrington 2014. Giving up beef will reduce carbon footprint more than cars, says expert. theguardian.com, 21 July.
46. The Vegan Society. Statistics. (Accessed 28 October 2020).
    Further reading: Sara Farr 2017. Why Reducing Meat Consumption Is the Easiest Step Everyone Can Take to Fix Our Broken Food System. *One Green Planet*.
    Further reading: Arthur Neslen 2018. Europe's meat and dairy production must halve by 2050, expert warns. theguardian.com, 15 September.
    Further reading: Damian Carrington 2018. Huge reduction in meat-eating 'essential' to avoid climate breakdown. theguardian.com, 10 October.
    Further reading: Rebecca Smithers 2018. Third of Britons have stopped or reduced eating meat – report. theguardian.com, 1 November.
47. Dave Goulson 2020. A lot to learn. *Resurgence & Ecologist*, Issue 318, January/February.
48. Jill L. Edmondson, et al. 2014. Urban cultivation in allotments maintains soil qualities adversely affected by

conventional agriculture. *Journal of Applied Ecology* volume 51, issue 4, August, 880-889.

49. Sara Burrows 2020. Nearly 40% of Russia's Food Still Comes From Small, Family Gardens. returntonow.net, February 13.
50. Barbara Harwood 2010, op. cit.
    Recommended book about water: Alick Bartholomew 2010. *The Spiritual Life of Water: Its Power and Purpose*. South Paris, ME: Park Street Press.
51. FAO 2018. The State of World Fisheries and Aquaculture 2018 – Meeting the sustainable development goals. Rome. Licence: CC BY-NC-SA 3.0 IGO.
52. D. Pauly and D. Zeller 2016. Catch reconstructions reveal that global marine fisheries catches are higher than reported and declining. *Nat. Commun.* 7:10244.
53. Marc Bekoff 2014. Fish Are Sentient and Emotional Beings and Clearly Feel Pain. psychologytoday.com, June 19.
54. Carl Safina 2018. Are we wrong to assume fish can't feel pain? theguardian.com, 30 October.
55. ISAAA 2018. Biotech Country, Facts and Trends: Brazil (2017). International Service for the Acquisition of Agri-biotech Applications.
56. Nicolai Kwasniewski 2018. Wie der Urwald für deutsches Fleisch gerodet wird. spiegel.de, 26.03.
57. FAO. How to Feed the World in 2050.
58. Quoted in: Arthur Neslen 2018. European parliament approves curbs on use of antibiotics on farm animals. theguardian.com, 25 October.

Chapter 13: Energy and "Progress"

1. Rob Mielcarski 2015. un-Denial Manifesto: Energy and Denial. un-Denial, November 12.
2. William E. Rees 2014. *Why Degrowth?* 2014 Degrowth Event Series, Vancouver BC, April 16, 2014. resilience.org, April

24, 2014. Min 10.09.

3. Michael Novack quote: personal communications, Ecocentric Alliance.
4. All Gorz quotes: André Gorz 1983 [1980]. *Ecology as Politics*. London: Pluto Press. 7-13.
5. Ivan Illich, quoted in Gorz 1983, op. cit. 7-8.
6. John Harris 2018. 'We'll have space bots with lasers, killing plants': the rise of the robot farmer. theguardian.com, 20 October.
7. Xiaowei Wang 2020. Behind China's 'pork miracle': how technology is transforming rural hog farming. theguardian.com, 8 October.
8. Daniela Siebert 2018. Studie zur Digitalisierung der Landwirtschaft. DLF Umwelt & Verbraucher, 10.10.
9. Duncan Clark and Mike Berners-Lee 2010. What's the carbon footprint of … the internet? theguardian.com, 12 August.
10. Climate Home News 2017. 'Tsunami of data' could consume one fifth of global electricity by 2025. theguardian.com, 11 December.
11. Christopher Helman 2016. Berkeley Lab: It Takes 70 Billion Kilowatt Hours A Year To Run The Internet. forbes.com, June 28.
12. Climate Home News 2017, op. cit.
13. Susanne Köhler, Massimo Pizzol 2019. Life Cycle Assessment of Bitcoin Mining. *Environmental Science & Technology* 53 (23): 13598–13606.
14. Chris Baraniuk 2019. Bitcoin's global energy use 'equals Switzerland'. BBC News, 3 July.
15. Camilo Mora, et al. 2018. Bitcoin emissions alone could push global warming above 2°C. *Nature Climate Change* 8 (11): 931-933.
16. Susanne Köhler, Massimo Pizzol 2019, op. cit.
17. Christian Stoll, et al. 2019. The Carbon Footprint of Bitcoin.

Joule 3 (7): 1647-1661.

18. Jeremy Rifkin 2019. *The Green New Deal: Why the Fossil Fuel Civilization Will Collapse by 2028, and the Bold Economic Plan to Save Life on Earth*. New York: St. Martin's Press.
19. Jillian Ambrose 2020. Oil prices dip below zero as producers forced to pay to dispose of excess. theguardian.com, 20 April.
20. Yves Smith 2020. With Bankruptcies Mounting, Faltering Oil and Gas Firms Are Leaving a Multi-Billion Dollar Cleanup Bill to the Public. nakedcapitalism.com, October 18.

    Pippa Stevens 2020. Shale industry will be rocked by $300 billion in losses and a wave of bankruptcies, Deloitte says. cnbc.com, June 22.
21. Max Wakefield 2020. Let's enjoy some good climate news: the block on UK onshore wind farms is no more. theguardian.com, 3 March.
22. Adam Morton 2020. Wind and solar plants will soon be cheaper than coal in all big markets around world, analysis finds. theguardian.com, 12 March.
23. Key World Energy Statistics 2020. Statistics report – August 2020.
24. Southern Environmental Law Center 2019. Fact Sheet: New Report Shows Wood Pellets from Drax's U.S. Mills Increase Carbon Emissions During the Timeframe Necessary to Address Climate Change.
25. Stand.Earth 2020. Risky business: Canada props up wood pellet export as a false climate solution. April 23.
26. Prof. John Beddington, et al. 2017. EU must not burn the world's forests for 'renewable' energy. Letters, theguardian.com, 14 December.
27. Partnership for Policy Integrity 2018. Letter from scientists to the EU Parliament regarding forest biomass (updated January 14, 2018).

28. Matt Hongoltz-Hetling 2020. US demand for clean energy destroying Canada's environment, indigenous peoples say. theguardian.com, 22 June.
29. The best possible introduction to Degrowth (also available on YouTube): William E. Rees 2014. *Why Degrowth?* 2014 Degrowth Event Series, Vancouver BC, April 16, 2014. resilience.org, April 24. Min 6.16.
30. All quotes in the Degrowth section from: William E. Rees 2020. Ecological economics for humanity's plague phase. *Ecological Economics* volume 169, March, 106519.
31. All quotes in this paragraph: William E. Rees 2020, op. cit.
32. *The Guardian,* Letters 2018. The EU needs a stability and wellbeing pact, not more growth. theguardian.com, 16 September.
33. WeMove 2018. Europe, It's Time to End the Growth Dependency.
34. Statista. Number of smartphone users worldwide from 2014 to 2020.
35. André Gorz 1983, op. cit. 69-74.
36. John Harris 2018. Our phones and gadgets are now endangering the planet. theguardian.com, 17 July.
37. Carbon Tracker 2020. Coal developers risk $600 billion as renewables outcompete worldwide. carbontracker.org, 12 March.
38. MP Mills 2019. The New Energy Economy: An Exercise in Magical Thinking. Manhattan Institute.
39. BP 2019. BP Statistical Review of World Energy 2019. British Petroleum, June 2019. (Accessed 28 October 2020).
40. Jillian Ambrose and Niko Kommenda 2020. Britain breaks record for coal-free power generation. theguardian.com, 28 April.
41. Jillian Ambrose 2020. UK electricity coal free for first month ever. theguardian.com, 2 June 2020.

Chapter 14: Climate Disruption

1. Guterres at the launch of the WMO Statement on the State of the Global Climate 2019, quoted in Damian Carrington 2020b. Climate emergency: global action is 'way off track' says UN head. theguardian.com, 10 March.
2. Zeke Hausfather 2020. State of the climate: 2020 on course to be warmest year on record. State of the Climate, *Carbon Brief*, 23.10.
3. Damian Carrington 2020b, op. cit.
4. John Vidal and Adam Vaughan 2012. Arctic sea ice shrinks to smallest extent ever recorded. 14 September.
5. Tim Flannery 2018. The Big Melt. *The New York Review of Books*, August 16.
6. Craig Welch 2018. Exclusive: Some Arctic Ground No Longer Freezing—Even in Winter. *National Geographic*, August 20.
7. Damian Carrington 2020b, op. cit.
8. Damian Carrington 2020c. Polar ice caps melting six times faster than in 1990s. theguardian.com, 11 March.
9. Ingo Sasgen, et al. 2020. Return to rapid ice loss in Greenland and record loss in 2019 detected by the GRACE-FO satellites. *Communications Earth & Environment* volume 1: 8.
10. Antje Boetius (interview by Christoph Seidler) 2019. Die wahren Herrscher der Welt. spiegel.de, 25.06.
11. Damian Carrington 2020a. Climate emergency: 2019 was second hottest year on record. theguardian.com, 15 January.
12. Andrea Thompson 2020. Heat and Humidity Are Already Reaching the Limits of Human Tolerance. scientificamerican.com, May 8.
13. Chi Xu, Timothy A. Kohler, Timothy M. Lenton, Jens-Christian Svenning, and Marten Scheffer 2020. Future of the human climate niche. *PNAS*, May 26, 2020 117 (21)

11350-11355; first published May 4.
14. Lisa Cox 2018. Australia heatwave to break Christmas weather records with temperatures up to 47C forecast. theguardian.com, 24 December.
15. The World 2020. 1 billion animals have died in Australian bushfires, ecologist estimates. pri.org, January 07.
Also: The University of Sydney 2020. More than one billion animals killed in Australian bushfires. sydney.edu.au, 8 January.
16. Joëlle Gergis 2020. We are seeing the very worst of our scientific predictions come to pass in these bushfires. theguardian.com, 3 January.
17. Reuters 2020. Fires in Amazon forest rose 30% in 2019. uk.reuters.com, January 9.
18. Reuters in Brasîlia 2020. Brazil's Amazon rainforest suffers worst fires in a decade. theguardian.com, 1 October.
19. Uki Goñi, Sam Cowie, William Costa 2020. 'Total destruction': why fires are tearing across South America. theguardian.com, 9 October.
20. Sarah Kaplan 2018. How humans have made wildfires worse. washingtonpost.com, August 14.
21. Oliver Milman, Vivian Ho 2020. California wildfires spawn first 'gigafire' in modern history. theguardian.com, 6 October.
22. Maanvi Singh 2020. 'Unprecedented': the US west's wildfire catastrophe explained. theguardian.com, 12 September.
23. Andrew Ciavarella, et al. 2020. Siberian heatwave of 2020 almost impossible without climate change. worldweatherattribution.org, 15 July.
24. Zicheng Yu, et al. 2010. Global peatland dynamics since the Last Glacial Maximum. *Geophysical Research Letters* volume 37, L13402, 9 July.
25. Richard Stone 2020. Siberia's 'gateway to the underworld' grows as record heat wave thaws permafrost. sciencemag.

org, July 28.

26. Antonio Cascais 2019. Amazon versus Africa forest fires: Is the world really ablaze? *DW*, 30 August.
27. Andrew J. Dowdy 2018. Climatological Variability of Fire Weather in Australia. *J. Appl. Meteor. Climatol.* (2018) 57 (2): 221-234.
    YouTube: "Nerilie Abram 2020. Animated history of rainfall and maximum temperature across southern Australia since 1910".
28. Nerilie J. Abram, Robert Mulvaney, et al. 2014. Evolution of the Southern Annular Mode during the past millennium. *Nature Climate Change* 4, 564-569 (2014).
29. Giovanni Di Virgilio, Jason P. Evans, et al. 2019. Climate Change Increases the Potential for Extreme Wildfires. *Geophysical Research Letters* volume 46, issue 14, 28 July 2019, 8517-8526.
    Andrew J. Dowdy, Hua Ye, Acacia Pepler, et al. 2019. Future changes in extreme weather and pyroconvection risk factors for Australian wildfires. *Scientific Reports* 9:10073.
30. Graham Readfearn 2020. Explainer: what are the underlying causes of Australia's shocking bushfire season? theguardian.com, 12 January.
31. Scott L. Stephens, Robert E. Martin, Nicholas E. Clinton 2007. Prehistoric fire area and emissions from California's forests, woodlands, shrublands, and grasslands. *Science Direct*, Forest Ecology and Management.
32. Sarah Kaplan 2018, op. cit.
33. CAL FIRE 2020. https://www.fire.ca.gov/
34. Andrew Ciavarella, et al. 2020, op. cit.
35. Damian Carrington 2018. 'Brutal news': global carbon emissions jump to all-time high in 2018. theguardian.com, 5 December.
36. Zeke Hausfather 2019. Analysis: Global fossil-fuel

emissions up 0.6% in 2019 due to China. *Carbon Brief*, 4 December.

37. Yangyang Xu, Veerabhadran Ramanathan and David G. Victor 2018. Global warming will happen faster than we think. *Nature* volume 564, 30-32, 6 December.
38. Sandy Irvine 2020. Covid-19 and a chance for sustainability. Sandy Irvine's green blog, March 31.
39. Damian Carrington 2020. Fifth of countries at risk of ecosystem collapse, analysis finds. theguardian.com, 12 October.
40. Migration Data Portal 2020. Environmental Migration.
41. Benjamin Franta 2018. Shell and Exxon's secret 1980s climate change warnings. theguardian.com, 19 September.
42. Franta 2018, op. cit.
43. Franta 2018, op. cit.
44. Syukuro Manabe 1970. *The Dependence of Atmospheric Temperature on the Concentration of Carbon Dioxide.*
    J. Hansen, D. Johnson, A. Lacis, S. Lebedeff, P. Lee, D. Rind, G. Russell 1981. Climate Impact of Increasing Atmospheric Carbon Dioxide. *Science*, 28 August 1981: Vol. 213, Issue 4511, 957-966.
    J. Hansen, I. Fung, A. Lacis, et al. 1988. Global climate changes as forecast by Goddard Institute for Space Studies three-dimensional model. *AGU Advancing Earth and Space Science*, 20 August.
45. John H. Cushman Jr. 2018. Shell Knew Fossil Fuels Created Climate Change Risks Back in 1980s, Internal Documents Show. *Inside Climate News*, April 5.
    Also: Banerjee, et al. 2015, op. cit.; Marco Evers 2018, op. cit.
46. InfluenceMap 2016. An investor enquiry: how much big oil spends on climate lobbying. InfluenceMap, April.
47. *Inside Climate News*: https://insideclimatenews.org/
    Also check for updates at Climate Files. http://www.

climatefiles.com/

48. *Inside Climate News*. https://www.pulitzer.org/finalists/insideclimate-news
49. Benjamin Franta 2018, op. cit.
50. Katherine Bagley 2015. 350,000 Sign Petition Asking for Federal Probe of Exxon. *Inside Climate News*, November 19.
51. Gordon MacDonald's How to Wreck the Environment. In Nathaniel Rich 2018. Losing Earth: The Decade We Almost Stopped Climate Change. nytimes.com, August 1.
52. Dana Nuccitelli 2013. Survey finds 97% of climate science papers agree warming is man-made. theguardian.com, 16 May.
53. Chris Mooney and Brady Dennis 2018. The world has just over a decade to get climate change under control, U.N. scientists say. washingtonpost.com, October 7.
54. For example, Germany: Christoph Seidler 2018. Die Welt gerät aus den Fugen – fragt sich nur, wie sehr. spiegel.de, 08.10.
55. 207 environmental defenders have been killed in 2017 while protecting their community's land or natural resources.
    Also: Jonathan Watts 2018. Almost four environmental defenders a week killed in 2017. theguardian.com, 2 February.
    Also: Mary Colwell 2018. The bloody truth about conservation: we need to talk about killing. theguardian.com, 28 May.
    Also: Jonathan Watts and John Vidal 2017. Environmental defenders being killed in record numbers globally, new research reveals. theguardian.com, 13 July.
56. Naomi Klein 2019. *On Fire: The (Burning) Case for a Green New Deal*. Allen Lane. 70.
57. Suzanne Goldenberg 2013. Climate research nearly unanimous on human causes, survey finds. theguardian.com, 16 May.

58. The Consensus Project. theconsensusproject.com
The Guardian has a category called "the 97%": https://www.theguardian.com/environment/climate-consensus-97-per-cent
59. *Skeptical Science.* https://skepticalscience.com/graphics.php?c=6
60. *Daily CO_2.* https://www.co2.earth/daily-co2
61. Quoted in Damian Carrington 2019. Climate-heating greenhouse gases hit new high, UN reports. theguardian.com, 25 November.
62. Union of Concerned Scientists: https://www.ucsusa.org/global-warming#.XDHwxYW9Bjc
Recommended: WJ Ripple, C. Wolf, TM Newsome, et al., and 15,364 scientist signatories from 184 countries, 2017. World Scientists' Warning to Humanity: A Second Notice. *BioScience* Volume 67, Issue 12, December, 1026-8.
63. A lot of suggestions are featured in Ernst Ulrich von Weizsaecker, Anders Wijkman, et al. 2018. *Come On! Capitalism, Short-termism, Population and the Destruction of the Planet. A Report to the Club of Rome.* New York: Springer.
64. – US traffic emissions: Fast Facts on Transportation Greenhouse Gas Emissions.
– UK traffic emissions: Josh Gabbatiss 2018. Transport becomes most polluting UK sector as greenhouse gas emissions drop overall. independent.co.uk, 06 February.
65. B. Lugschitz, M. Bruckner and S. Giljum 2011. *Europe's global land demand. A study on the actual land embodied in European imports and exports of agricultural and forestry products.* Vienna: Sustainable Europe Research Institute.
66. World Resources Institute. World Greenhouse Gas Emissions: 2016.
67. Frances Seymour and Jonah Busch. *Why Forests? Why Now? The Science, Economics, and Politics of Tropical Forests and Climate Change.* Washington DC: Center for Global

Development. 40.
68. UNEP 2017. The Emissions Gap Report 2017. United Nations Environment Programme (UNEP), Nairobi. xiv.
69. FAO 2016. The State of Food and Agriculture: Climate Change, Agriculture and Food Security. Rome.
70. Tim Flannery 2011. *Here on Earth – A Twin Biography of the Planet and the Human Race*. London/New York/Toronto: Penguin. 260, 263.
71. Recommended: Damian Carrington 2018. Global food system is broken, say world's science academies. theguardian.com, 28 November.
72. J. Poore, T. Nemecek 2018. Reducing food's environmental impacts through producers and consumers. *Science*, 01 June 2018: Vol. 360, Issue 6392, 987-992.
73. Helen Harwatt 2018. Including animal to plant protein shifts in climate change mitigation policy: a proposed three-step strategy. *Climate Policy*.
74. Maneka Gandhi 1990. Save the Trees, Don't Eat Meat. *The Illustrated Weekly of India*, 11.11.1990.
75. FAO. By the numbers: GHG emissions by livestock. FAO Key facts and findings. (Accessed 28 October 2020). After: FAO 2013. *Tackling Climate Change Through Livestock: A global assessment of emissions and mitigation opportunities*. Rome.
76. IMO 2015. Third IMO GHG Study 2014: Executive Summary and Final Report.
77. David Pimentel and Marcia Pimentel 1979. *Food, Energy, and Society* (Resource and Environmental Sciences Series). Hoboken, NJ: Wiley.
78. Harwatt, William J. Ripple, et al. 2019. Scientists call for renewed Paris pledges to transform agriculture. *The Lancet – Planetary Health*, December 11.
79. Rebecca Smithers 2018. Third of Britons have stopped or reduced eating meat – report. theguardian.com, 1

November.
80. Cornelia Rumpel 2018. Put more carbon in soils to meet Paris climate pledges. *Nature* volume 564, 32-34, 6 December.
81. Charles Eisenstein 2015. We need regenerative farming, not geoengineering. theguardian.com, 9 March 2015.
82. John Vidal 2008. True scale of CO2 emissions from shipping revealed. theguardian.com, 13 February.
83. – 2.6%: IMO 2015. Third IMO GHG Study 2014: Executive Summary and Final Report.
– 3%: Rahul Mudliar 2020. UN and Global banks focusing on green shipping to cap Carbon Dioxide Emissions from Global Shipping. *SG Analytics*, January 2.
84. John Vidal 2009. Health risks of shipping pollution have been 'underestimated'. theguardian.com, 9 April.
85. Mike Scott 2014. Sustainable shipping is making waves. theguardian.com, 1 August.
86. UCAR NCAR 2015.
87. Faig Abbasov 2019. One corporation to pollute them all. transportenvironment.org, June 4.
88. Mark Matousek 2019. Carnival reportedly dumped over 500,000 gallons of sewage and 11,000 gallons of food waste improperly in the year after it admitted to illegally releasing oil into the ocean. businessinsider.com, April 17.
89. Friends of the Earth 2019. Cruise Ships.
90. Cruise Lines International Association (CLIA) 2019. Cruise Trends & Industry Outlook.
91. ibid.
92. Annelies Wilder-Smith, Professor of Emerging Infectious Diseases at the London School of Hygiene and Tropical Medicine, in: Alice Hancock 2020, 'Floating Petri dishes': The 2020s were meant to be a boom decade for cruises—then COVID-19 hit them like a tidal wave. *Financial Times*, June 10.
93. taz 2012. Steckdose für Kreuzfahrtschiffe. taz.de, 9.8.2012.

94. German test in: auto motor sport 2020. Öko oder Mogelpackung? 7 Fragen zum Plug-In-Hybrid - Bloch erklärt #86.
95. planet e 2018. Der wahre Preis der Elektroautos. ZDF, 09.09.2018.
Also: Dirk Blijweert 2018. Scientists reveal the hidden costs of cobalt mining in DR Congo. phys.org, September 21.
96. Sandy Irvine's green blog. https://sandyirvineblog.wordpress.com
97. Adam Vaughan 2018. Use excess wind and solar power to produce hydrogen – report. theguardian.com, 9 May 2018.
98. Recommended: Jonah M. Kessel, Hiroko Tabuchi 2019. Texas Methane Super Emitters. nytimes.com, December 12.
99. Damian Carrington 2019. Climate-heating greenhouse gases hit new high, UN reports. theguardian.com, 25 November.
100. Benjamin Hmiel, et al. 2020. Preindustrial 14CH4 indicates greater anthropogenic fossil CH4 emissions. *Nature* volume 578, 409-412 (2020).
101. Ramón A. Alvarez, et al. 2018. Assessment of methane emissions from the U.S. oil and gas supply chain. *Science,* 13 July 2018, Vol. 361, Issue 6398, 186-188.
102. RB Jackson, et al. 2020. Increasing anthropogenic methane emissions arise equally from agricultural and fossil fuel sources. *Environmental Research Letters* volume 15, No. 7.
Also: Marielle Saunois, et al. 2020. The Global Methane Budget 2000–2017. *Earth Syst. Sci. Data* 12, 1561-1623.
103. Will Steffen, et al. 2018. Trajectories of the Earth System in the Anthropocene. *PNAS,* August 14, 2018 115 (33) 8252-8259.
104. Tim Lenton quoted in Damian Carrington 2019. Climate emergency: world 'may have crossed tipping points'. theguardian.com, 27 November.
105. L. Beaufort, et al. 2011. Sensitivity of coccolithophores to

carbonate chemistry and ocean acidification. *Nature* 476 (7358): 80-3, doi:10.1038/nature10295, PMID 21814280. https://www.nature.com/articles/nature10295#ref8

106. Simon L. Lewis, et al. 2011. The 2010 Amazon Drought. *Science*, 04 February, Vol. 331, Issue 6017, 554.
   Fiona Harvey 2020. Tropical forests losing their ability to absorb carbon, study finds. theguardian.com, 4 March.
107. Carlos Nobre, a climate scientist at the University of São Paolo in Brazil, in Quirin Schiermeier 2019. Eat less meat: UN climate-change report calls for change to human diet. *Nature* 572, 291-292 (2019), 08 August 2019.
108. Timothy M. Lenton, et al. 2020. Climate tipping points — too risky to bet against. 27 November 2019, Correction 09 April 2020. *Nature* volume 575, 592-595.
109. Gregory S. Cooper, Simon Willcock & John A. Dearing 2020. Regime shifts occur disproportionately faster in larger ecosystems. *Nature Communications* volume 11, Article 1175.
110. Asshoff, et al. 2006. Growth and phenology of mature temperate forest trees in elevated CO_2. *Global Change Biology* 12, 848-861.
111. Dahr Jamail 2018. How Feedback Loops Are Driving Runaway Climate Change. *Truthout*, October 1.
112. Damian Carrington 2018. Avoid Gulf stream disruption at all costs, scientists warn. theguardian.com, 13 April.
113. ME Mann, et al. 2017. Influence of anthropogenic climate change on planetary wave resonance and extreme weather events. *Scientific Reports* 7, 27 March.
   Also: D. Coumou, et al. 2018. The influence of arctic amplification on mid-latitude summer circulation. *Nature Communications* 9, 20 August.
114. ME Mann, et al. 2018. Projected changes in persistent extreme summer weather events: The role of quasi-resonant amplification. *Science Advances* 4, no. 10, 31 October.

115. Natalia Shakhova, et al. 2017. Current rates and mechanisms of subsea permafrost degradation in the East Siberian Arctic Shelf. *Nature Communications* 8, 15872. 22 June. Also: N. Shakhova, I. Semiletov, I. Leifer, A. Salyuk, P. Rekant, D. Kosmach 2010. Geochemical and geophysical evidence of methane release over the East Siberian Arctic Shelf. *Journal of Geophysical Research*, 07 August.
116. Natalia Shakhova 2012. Interview at the European Geophysical Union in Vienna, 2012. On YouTube: "Methane Hydrates – Extended Interview Extracts With Natalia Shakhova".
117. European Parliament News 2020. COVID-19: MEPs call for massive recovery package and Coronavirus Solidarity Fund. Press Release, 17-04-2020.
118. Stan Cox 2020. Fair Enough. resilience.org, May 4, 2020.
119. Cornelia Rumpel 2018. Put more carbon in soils to meet Paris climate pledges. *Nature* volume 564, 32-34, 6 December 2018.
120. Dominic Bliss 2020. Universal Basic Income is gathering support. Has it ever worked – and could it work in the UK? nationalgeographic.co.uk, 20 May.
121. For example: Jeremy Rifkin 2019. *The Green New Deal: Why the Fossil Fuel Civilization Will Collapse by 2028, and the Bold Economic Plan to Save Life on Earth*. New York: St. Martin's Press.
122. UN Sustainable Development Goals: https://sustainabledevelopment.un.org/sdgs
123. UN Sustainable Development Goals 2020. Transforming our world: the 2030 Agenda for Sustainable Development.
124. Gilbert Rist 2019. *The History of Development: From Western Origins to Global Faith*. London: Zedbooks. 3.
125. Rist 2019, op. cit. 6.
126. ibid.
127. World Commission on Environment and Development

1988. *Our Common Future,* with an introduction by Gro Harlem Brundtland. London: Fontana Books.

128. *Our Common Future,* pp. 6, xii; quoted in Rist 2019, 183.
129. Rist 2019, 194.
130. NASA Global Climate Change 2018. Global Temperature. https://climate.nasa.gov/vital-signs/global-temperature/
131. Oliver Milman 2018b: Climate change 'will inflict substantial damages on US lives'. theguardian.com, 23 November.

    Also: Oliver Milman 2018c: World 'nowhere near on track' to avoid warming beyond 1.5C target. theguardian.com, 27 September.
132. Joachim Müller-Jung und Christian Schwägerl 2018. "Klimaschutz ist kein Wunschkonzert". faz.net, 01. Okt.
133. State of Alaska Department of Natural Resources, Geological & Geophysical Surveys 2020. Barry Arm Landslide and Tsunami Hazard.
134. Neela Banerjee, Lisa Song and David Hasemyer 2015. Exxon's Own Research Confirmed Fossil Fuels' Role in Global Warming Decades Ago. *ICN,* September 15.
135. Banerjee, et al. 2015, op. cit.
136. See the opening sequence in the Jeff Gibbs and Michael Moore documentary *Planet of the Humans.*
137. Justin Farrell 2015. Corporate funding and ideological polarization about climate change. *PNAS,* November 23, referring to a Harvard study.
138. Dana Nuccitelli 2015. Two-faced Exxon: the misinformation campaign against its own scientists. theguardian.com, 25 November.
139. InfluenceMap 2016, op. cit.

    Further reading about Exxon's climate PR: Geoffrey Supran and Naomi Oreskes 2017. What Exxon Mobil Didn't Say About Climate Change. nytimes.com, August 22.
140. Dana Nuccitelli 2015, op. cit.

Also: Suzanne Goldenberg 2015. Work of prominent climate change denier was funded by energy industry. theguardian.com, 21 February.

141. Suzanne Goldenberg and Helena Bengtsson 2016. Biggest US coal company funded dozens of groups questioning climate change. theguardian.com, 13 June.
Lee Fang 2015. Attorney Hounding Climate Scientists Is Covertly Funded By Coal Industry. *The Intercept*, August 25.
142. Climate-denial extremists such as Richard Lindzen, Willie Soon, Richard Berman, and Roy Spencer. In: Suzanne Goldenberg and Helena Bengtsson 2016, op. cit.
143. Lee Fang 2015, op. cit.
144. Bagley 2015, op. cit.
145. Dana Nuccitelli 2015, op. cit.
146. John H. Cushman Jr. 2018, op. cit.
147. InfluenceMap 2016. An investor enquiry: how much big oil spends on climate lobbying. InfluenceMap, April 2016.
148. David Hasemyer 2018. Fossil Fuels on Trial: Where the Major Climate Change Lawsuits Stand Today. *ICN*, November 6.
149. Georg Ehring 2018. Klagen für mehr Klimaschutz: Weltweit immer mehr Gerichtsverfahren. DLF, Umwelt & Verbraucher, 6.12.
150. Dana Nuccitelli 2015, op. cit. Referring to Bill McKibben 2015. Imagine if Exxon had told the truth on climate change. theguardian.com, 28 October.
151. Nathaniel Rich 2018, op. cit.
152. ibid.
153. Dana Nuccitelli 2013, op. cit.
154. CBS 2018. Sexual assault victims confront Jeff Flake over support for Kavanaugh – video. theguardian.com, 28 September.
155. Juliet Eilperin, Brady Dennis, and Chris Mooney 2018.

Trump administration sees a 7-degree rise in global temperatures by 2100. *The Washington Post*, September 28.

156. Coral Davenport and Kendra Pierre-Louis 2018. U.S. Climate Report Warns of Damaged Environment and Shrinking Economy. nytimes.com, November 23.
157. JISAO 2017. Pacific Decadal Oscillation (PDO). February.
Also: Benjamin D. Santer, et al. 2014. Volcanic contribution to decadal changes in tropospheric temperature. *Nature Geoscience* volume 7, 185-189, 23 February.
Also: Matthew H. England, et al. 2014. Recent intensification of wind-driven circulation in the Pacific and the ongoing warming hiatus. *Nature Climate Change* volume 4, 222-227, 23 February.
158. Sarah McFarlane and Jonathan Saul 2014. Food importers shift from dry bulk cargo ships to containers. in.reuters.com, February 14.
159. John Vidal 2008, op. cit.
160. Brandon Graver, Kevin Zhang, Dan Rutherford 2018. CO2 emissions from commercial aviation, 2018. The International Council on Clean Transportation, September 2019.
161. IMO International Maritime Organization 2014. Greenhouse Gas Emissions.
162. The International Maritime Organisation's MARPOL Annex VI regulation, which reduces sulfur emissions in key Emission Control Areas (ECAs): the North Sea, Baltic Sea, most of the US and Canadian coastline, and the US Caribbean.
163. BBC World Service 2017. Big Polluters: Ships v Cars. 2 October.
164. John Vidal 2009, op. cit.
165. Cruise Lines International Association (CLIA) 2019, op. cit.
166. B. Wolber, A. Papathanassis, M. Vogel 2012. *The Business and Management of Ocean Cruises*. CABI.

167. Prof. D. Koehler 2018: (German radio interview by) Stefan Michel 2018. Gefährlichkeit von Stickstoffdioxid bleibt umstritten. DLF, Umwelt & Verbraucher, 14.12.
168. Shakuntala Makhijani 2014. Fossil fuel exploration subsidies: United States. odi.org, November.
169. Shakuntala Makhijani 2014. Fossil fuel exploration subsidies: United Kingdom. odi.org, November.
170. ibid. 70-72.

Part III: The Human Interface

1. Tijjani Muhammad-Bande 2020. UN International Mother Earth Day 22 April Message. https://www.un.org/en/observances/earth-day/message

Chapter 15: Why Does So Little Happen, and So Slowly?

1. Wade Davis 2018. On Ecological Amnesia. *The Tyee*, 8 November.
2. Haydn Washington and Helen Kopnina 2018. The insanity of endless growth. *The Ecological Citizen* 2: 57-63.
3. – James Randerson 2006. World's richest 1% own 40% of all wealth, UN report discovers. theguardian.com, 6 December.
   – OXFAM International 2018. Richest 1 percent bagged 82 percent of wealth created last year - poorest half of humanity got nothing.
   – Graeme Wearden 2014. Oxfam: 85 richest people as wealthy as poorest half of the world. theguardian.com, 20 January.
   – OXFAM International 2017. Just 8 men own same wealth as half the world.
4. Washington and Kopnina 2018, op. cit.
5. Jameson quote: "It seems to be easier for us today to imagine the thoroughgoing deterioration of the earth and of nature

than the breakdown of late capitalism." Fredric Jameson 1994. *The Seeds of Time*. Columbia University Press.

6. Georg Diez 2018. Unser Lebensstil muss verhandelbar sein. spon, 12.08.
7. Quoted in: Georg Diez 2018, op. cit.
8. Naomi Klein 2019. *On Fire: The (Burning) Case for a Green New Deal*. Allen Lane. 149.
9. Haydn Washington and Helen Kopnina 2018. The insanity of endless growth. *The Ecological Citizen* 2: 57-63.
10. Gabriel Millar 2001. *The Saving Flame*. Five Seasons and Tumbled Stone Press.
11. Washington and Kopnina 2018, op. cit.
12. Dana Nuccitelli 2018. The Trump administration has entered Stage 5 climate denial. theguardian.com, 8 October.
13. Quoted in: Dana Nuccitelli 2018, op. cit.
14. Obama on September 7, 2018 in a speech at the University of Illinois, quoted in: Peter Baker 2018. Obama Lashes Trump in Debut 2018 Speech. President's Response: 'I Fell Asleep.' nytimes.com, September 7.
15. Excerpt from Jay Ramsay 2008. *Anamnesis* 11. London: The Lotus Foundation.
16. Arwen Long, Michael Platt 2005. Decision Making: The Virtue of Patience in Primates. *Current Biology* Volume 15, Issue 21, 8 November 2005, R874-R876.
17. William E. Rees 2020. Ecological economics for humanity's plague phase. *Ecological Economics* volume 169, March 2020, 106519.
18. Hoda Baraka, Muslim and 350.org's global communications director, quoted in Rebecca Solnit 2019. Why climate action is the antithesis of white supremacy. theguardian.com, 19 March.
19. Will Steffen, et al. 2018. Trajectories of the Earth System in the Anthropocene. *PNAS*, August 14, 2018 115 (33) 8252-8259.

20. Quoted in: Tobias Haberkorn 2018. Die Sintflut kommt. zeit.de, 4. November.
21. The Climate Mobilization 2017. The Climate Mobilization Begins in Los Angeles! theclimatemobilization.org, October 15.
22. Alison Green and Molly Scott Cato 2018. Facts about our ecological crisis are incontrovertible. We must take action. theguardian.com, 26 October.
23. WJ Ripple, C. Wolf, TM Newsome, et al., and 15,364 scientist signatories from 184 countries, 2017. World Scientists' Warning to Humanity: A Second Notice. *BioScience* Volume 67, Issue 12, December 2017, 1026-8.
24. Quoted in: David Crouch 2018. The Swedish 15-year-old who's cutting class to fight the climate crisis. theguardian.com, 1 September.
25. Greta Thunberg's UN speech (COP24) in December 2018: YouTube: "Greta Thunberg's COP24 speech in Katowice 2018".
    Also, on Vimeo: Greta Thunberg – full speech from COP24.
26. Demo posters from the Australian climate school strikes: Guardian staff 2018. Schools climate strike: the best protest banners and posters. theguardian.com, 30 November.
    And more posters: ABC News 2018. Students strike for climate change protests, defying calls to stay in school. abc.net.au, 30 November.
27. Zhou 2018, op. cit.
28. Damian Carrington 2018. 'Our leaders are like children,' school strike founder tells climate summit. theguardian.com, 4 December.
29. Anna Ringstrom, Clement Rossignol 2019. Sweden's Thunberg demands climate action on day of global school strikes. reuters.com, May 24.
30. Karn Vohra, Alina Vodonos, et al. 2021. Global mortality from outdoor fine particle pollution generated by fossil

fuel combustion: Results from GEOS-Chem. *Environmental Research* Volume 195, April, 110754.

31. Juergen Habermas 1996. Ziviler Ungehorsam – Testfall für den demokratischen Rechtsstaat. In Juergen Habermas: *Die Neue Unübersichtlichkeit – Kleine politische Schriften*. Frankfurt/Main.
32. André Gorz 1983 [1980]. *Ecology as Politics*. London: Pluto Press. 7-13.
33. On German radio DLF, 11.10.2018.
34. Dana Nuccitelli 2013. The 5 stages of climate denial are on display ahead of the IPCC report. theguardian.com, 16 September.
35. Extinction Rebellion: https://rebellion.earth
    Also: Roger Hallam 2018. Extinction Rebellion Diary #1: So it has come to this? theecologist.org, 23rd August.
    Also: Damien Gayle 2018. Climate protesters glue hands to UK government building. theguardian.com, 12 November.
36. XR interview with George Monbiot, audio starting at 17:22, in Anushka Asthana, et al. 2018. The plastics conspiracy: who is to blame for the waste crisis? theguardian.com, 14 November.
37. Oregon State University video clip about *The Second Warning*.
    Also: CBC News about *The Second Warning*.
38. Greta Thunberg 2018. Sweden is not a Role Model. *Medium*, August 24.
39. Naaman Zhou 2018. Climate change strike: thousands of school students protest across Australia. theguardian.com, 30 November.
    Also: Australian Geographic 2018. In pictures: School Strike for Climate Action. australiangeographic.com.au, November 30.
40. Liza Selley, Linda Schuster, et al. 2020. Brake dust exposure exacerbates inflammation and transiently compromises

phagocytosis in macrophages. *Metallomics*, Vol 12, Issue 3: 371-386. March.

Manju Mehta, Lung-Chi Chen, et al. 2008. Particulate matter inhibits DNA repair and enhances mutagenesis. *Mutation Research*. Dec 8; 657(2): 116-121.

Michelle C Turner, Daniel Krewski, et al. 2011. Long-term ambient fine particulate matter air pollution and lung cancer in a large cohort of never-smokers. *Am J Respir Crit Care Med*. Dec 15; 184(12): 1374-81.

Chapter 16: Anthropocentrism

1. All Greer quotes: John Michael Greer 2017. The twilight of anthropocentrism. *The Ecological Citizen* 1: 75-82.
2. John Michael Greer 2017, op. cit.
3. Aldo Leopold quote in: Patrick Curry 2017. The Ecological Citizen: An impulse of life, for life. *The Ecological Citizen* 1: 5-9.
4. Stan Rowe 1994. Ecocentrism: the Chord that Harmonizes Humans and Earth. *The Trumpeter* 11(2): 106-107. EcoSpherics.
5. On "ecosystem services": George Monbiot 2018. The pricing of everything. *The Ecological Citizen* 2: 89-96.
   Also: George Wuerthner 2018. Anthropocene boosters and the attack on wilderness conservation. *The Ecological Citizen* 1: 161-6.
   Also: Sian Sullivan 2017. Noting some effects of fabricating 'nature' as 'natural capital'. *The Ecological Citizen* 1: 65-73.
6. George Wuerthner 2018. Anthropocene boosters and the attack on wilderness conservation. *The Ecological Citizen* 1: 161-6.
7. Jack D. Forbes 2008 [1992, 1979]. *Columbus and Other Cannibals*. New York: Seven Stories Press. All quotes from the Introduction and Chapter 2.
8. In: Forbes 2008, op. cit. 21-22.

9. Erin McKenna, Scott L. Pratt 2015. *American Philosophy: From Wounded Knee to the Present*. London/New York: Bloomsbury. 375.
10. Eileen Crist 2017. The affliction of human supremacy. *The Ecological Citizen* 1: 61-4.
11. Charles Darwin 1859. *On the Origin of Species*.
12. LR Croft 1989. *The Life and Death of Charles Darwin*. Chorley: Elmwood.
13. Recommended: Peter A. Levine 1997. *Waking the Tiger: Healing Trauma*. Berkeley: North Atlantic Books.
14. James C. Scott 2018. *Against the Grain: A Deep History of the Earliest States*. Yale University Press. 8.
15. ibid. 33.
16. ibid. 113-114.
17. ibid. 7.
18. Lisi Krall 2018. The economic legacy of the Holocene. *The Ecological Citizen* 2: 67-76.
19. Henry Beston 1928. *The Outermost House: A Year of Life on the Great Beach of Cape Cod*.
20. John C. Lilly quoted in: Paul Watson (n.y.). The Cetacean Brain and Hominid Perceptions of Cetacean Intelligence. Unpublished essay.
21. Paul Watson (n.y.), op. cit.
22. Sherri Mitchell quote: my transcript from her talk for the Friends of Penobscot Bay.
23. George Wuerthner 2018, op. cit.
24. James C Scott 2018, op. cit. 11.
25. Lisi Krall 2018, op. cit.

Chapter 17: The Ecocentric Worldview

1. Stan Rowe 1994. Ecocentrism and Traditional Ecological Knowledge. *EcoSpherics*. http://www.ecospherics.net/pages/Ro993tek_1.html
2. Stan Rowe 1994. Ecocentrism: the Chord that Harmonizes

Humans and Earth. *The Trumpeter* 11(2): 106-107. *EcoSpherics*. http://www.ecospherics.net/pages/RoweEcocentrism.html
3. Patrick Curry 2017. The Ecological Citizen: An impulse of life, for life. *The Ecological Citizen* 1: 5-9.
4. Ted Mosquin and Stan Rowe 2004. A Manifesto for Earth. *Biodiversity* 5 (1), September 2004. *EcoSpherics*. http://www.ecospherics.net/
5. ibid.

Chapter 18: A Budding Future

1. Coyote Alberto Ruz Buenfil 2017. Enacting the wisdom of Chief Seattle today in Latin America. *The Ecological Citizen* 1: 55-9.
2. John Mohawk 2010. *Thinking in Indian: A John Mohawk Reader*. Golden, CO: Fulcrum.
3. Coyote Alberto Ruz Buenfil 2017, op. cit.
4. Marie-Lise Schläppy and Joe Gray 2017. Rights of nature: A report on a conference in Switzerland. *The Ecological Citizen* 1: 95-6.
5. Geneva Forum. http://www.osi-genevaforum.org/United-Nations-2016-December-15th-Rights-of-Nature-for-Peace-and-Sustainable
6. https://www.iucn.org/commissions/world-commission-environmental-law/wcel-resources/wcel-important-documentation/environmental-rule-law
7. Mumta Ito 2017. Towards a new paradigm for nature in the EU: A report on a meeting in Belgium. *The Ecological Citizen* 1: 97-8.
8. EcoHustler 2018. Top Shell bosses accused of climate ecocide in The Hague. ecohustler.com, December 7.
9. The amendment was voted on 20th January 2021 by the Plenary Session of the EP, and then finally adopted (459/62/163). Stop Ecocide 2021. Press Release: European parliament urges support for making ecocide an

international crime. January 21.

10. Stan Rowe 2000. An Earth-Based Ethic for Humanity. *EcoSpherics*. http://www.ecospherics.net/pages/RoweEarth Ethics.html
11. *Laudato Si*, the original: The Vatican 2015. Encyclical Letter *Laudato Si of the Holy Father Francis On Care for Our Common Home*. http://w2.vatican.va/content/francesco/en/encyclicals/documents/papa-francesco_20150524_enciclica-laudato-si.html
    Laudato Si, my summary: Fred Hageneder 2015. Pope calls for caring for the Earth. themeaningoftrees.com, June 2015. https://themeaningoftrees.com/pope-calls-for-caring-for-the-earth/
12. All ARC infos and quotes from: ARC: Faiths and Ecology. http://www.arcworld.org/arc_and_the_faiths.asp
    Also: ARC Downloads: http://www.arcworld.org/downloads.asp
13. Stan Rowe 2000. An Earth-Based Ethic for Humanity. *EcoSpherics*. http://www.ecospherics.net/pages/RoweEarth Ethics.html
14. Fritjof Capra 1996. *The Web of Life: A New Scientific Understanding of Living Systems*. New York and Toronto: Anchor Books, Doubleday. Quoted in: Rowe 2000, op. cit.
15. George Monbiot 2020. Coronavirus shows us it's time to rethink everything. Let's start with education. theguardian.com, 12 May 2020.
16. Neil Gaiman and Chris Riddell 2018. *Art Matters: Because Your Imagination Can Change the World*. London: Headline. Quoted in: Neil Gaiman and Chris Riddell on why we need libraries – an essay in pictures. theguardian.com, 6 September.
17. Harald Lesch 2018. "Die Menschheit schafft sich ab". SWR Tele-Akademie. Quote at 25:20. YouTube.
    Further reading about new economy: Shann Turnbull 2018.

A vision for an ecocentric society and how to get there. *The Ecological Citizen* 1: 141-2.

18. Haydn Washington 2018. Harmony – not 'theory'. *The Ecological Citizen* 1: 203-10.
19. Roselle Angwin. https://thewildways.co.uk/
20. Sherri Mitchell quotes: my transcript from her talk for the Friends of Penobscot Bay.
21. Recommended: Joe Gray 2017. Reasons for a reduction of humans' impact on the ecosphere. *The Ecological Citizen* 1: 17-18.
22. Alexander Lautensach 2018. Learning for biosphere security in a crowded, warming world. *The Ecological Citizen* 1: 171-8.

Chapter 19: Finding Hope, Courage, and Strength

1. Joe Gray 2018. Green fidelity and the grand finesse: Stepping stones to the 'Pacocene'. *The Ecological Citizen* 1: 121-9.
2. Patrick Curry 2004 [1997]. *Defending Middle-Earth – Tolkien: Myth and Modernity*. Boston/New York: Houghton Mifflin. 148.
3. Patrick Curry 2004, op. cit. 149.
4. Mary Robinson quote in: Rory Carroll: Mary Robinson on climate change: "Feeling 'This is too big for me' is no use to anybody". theguardian.com, 12 October 2018.
5. Fiona Harvey 2018. World must triple efforts or face catastrophic climate change, says UN. theguardian.com, 27 November 2018.
6. Gandalf quote in: Patrick Curry 2017. The Ecological Citizen: An impulse of life, for life. *The Ecological Citizen* 1: 5-9.
7. Patrick Curry 2017, op. cit.
8. Patrick Curry 2011. *Ecological Ethics: An introduction*. Cambridge (UK): Polity Press. 269.

9. Patrick Curry 2004, op. cit. 150.
10. Russell Means, quoted in: Paul Watson 2018. Interview with Captain Paul Watson. *The Ecological Citizen* 1: 152-3.
11. ibid.
12. Jonathan Franzen: *End of the End of the Earth,* quoted in: Sarah Crown: *The End of the End of the Earth* by Jonathan Franzen review – hope in an age of crisis. theguardian.com, 9 November 2018.
13. Jack D. Forbes 2008 [1992, 1979]. *Columbus and Other Cannibals*. New York: Seven Stories Press. 184.

Afterword

1. Sherri Mitchell (Weh'na Ha'mu' Kwasset) 2018. *Sacred Instructions: Indigenous Wisdom for Living Spirit-Based Change*. Berkeley (CA): North Atlantic Books.

Picture Credits

All artwork by DesignIsIdentity.com. Watercolor Figure 3 by Myla Twilley-Lilly (10).

Figure 6: Yinon M. Bar-On, Rob Phillips, Ron Milo 2018. The biomass distribution on Earth. PNAS, June 19, 2018, 115 (25) 6506-6511.

Figure 7: Andreas Kortenkamp, Michael Faust 2018. Regulate to reduce chemical mixture risk. *Science*, 20 July, vol. 361, issue 6399, 224–6.

Figure 8: Damian Carrington 2017. Want to fight climate change? Have fewer children. theguardian.com, 12 July.

Figure 9: D. Pauly and D. Zeller 2016. Catch reconstructions reveal that global marine fisheries catches are higher than reported and declining. *Nat. Commun.* 7:10244.

Figure 10: Yinon M. Bar-On, Rob Phillips, Ron Milo 2018. The biomass distribution on Earth. *PNAS*, June 19, 115 (25) 6506-6511.

Figure 11: J. Poore, T. Nemecek 2018. Reducing food's

environmental impacts through producers and consumers. *Science* 01 June: Vol. 360, Issue 6392, 987-992.

Figure 12: EST 2019. Soja – Globale Mengenverteilung und Mengenströme. EST, accessed June 2020.

Figure 13: Ingo Sasgen, et al. 2020. Return to rapid ice loss in Greenland and record loss in 2019 detected by the GRACE-FO satellites. *Communications Earth & Environment* volume 1: 8.

Figure 14: National Snow and Ice Data Center/Thomas Mote, University of Georgia. In: Jenessa Duncombe 2019. Greenland Ice Sheet Beats All-Time 1-Day Melt Record. *EOS*, 2 August 2019.

Figure 15: US Earth System Research Laboratory.

Figure 16: John Cook, Dana Nuccitelli, et al. 2013. Quantifying the consensus on anthropogenic global warming in the scientific literature. IOPscience, *Environmental Research Letters* Volume 8, Number 2.

Figure 17: Vostok ice core data. In: Stephan Harding 2009. *Animate Earth: Science, Intuition and Gaia,* 2nd ed. Cambridge (UK): Green Books.

Figure 18: UNEP 2020. The Emissions Gap Report 2020. United Nations Environment Programme (UNEP), Nairobi. 4.

Figure 19: UNEP 2019. The Emissions Gap Report 2019. United Nations Environment Programme (UNEP), Nairobi. xiv. Emissions Gap Report 2020. xiv.

Figure 20: Frances Seymour and Jonah Busch 2016. *Why Forests? Why Now? The Science, Economics, and Politics of Tropical Forests and Climate Change*. Washington DC: Center for Global Development. 40.

Figure 21: Peter Scarborough, et al. 2014. Dietary greenhouse gas emissions of meat-eaters, fish-eaters, vegetarians and vegans in the UK. *Climatic Change* volume 125, 179-192.

Figure 22: Robert McSweeney 2020. Scientists concerned by 'record high' global methane emissions. *Carbon Brief*, 14 July.

Marielle Saunois, et al. 2020. The Global Methane Budget 2000–2017. *Earth Syst. Sci. Data* 12, 1561-1623.

Figure 23: Sarah Fecht 2018. Sluggish Ocean Currents Caused European Heat Wave Some 12,000 Years Ago. *State of the Planet*, Earth Institute, Columbia University, 24 April.

Figure 24 + 25: ME Mann 2019. Klimawandel: Gefährlicher Wetterverstärker. *Spektrum der Wissenschaft* 7.19, 54-61. World map by Natalya Koltovskaya/shutterstock.com

Further Reading

Mike Berners-Lee 2019. *There Is No Planet B: A Handbook for the Make or Break Years*. Cambridge University Press.

Ernest Callenbach 2008: *Ecology – A Pocket Guide*. Berkeley, Los Angeles, London: University of California Press.

Eileen Crist and H. Bruce Rinker (eds.) 2010. *Gaia in Turmoil*. Cambridge, MA: The MIT Press.

Patrick Curry 2011. *Ecological Ethics: An introduction*. Cambridge, UK: Polity Press.

Patrick Curry 2019. *Enchantment: Wonder in Modern Life*. Edinburgh: Floris Books.

Christiana Figueres, Tom Rivett-Carnac 2020. *The Future We Choose: Surviving the Climate Crisis*. London: Manilla Press.

Stephan Harding 2009. *Animate Earth: Science, Intuition and Gaia*, 2nd edition. Cambridge, UK: Green Books.

Rob Hopkins 2019. *From What Is to What If*. White River Junction, VT: Chelsea Green Publishing.

Ernst Ulrich von Weizsaecker, Anders Wijkman, et al. 2017. *Come On! Capitalism, Short-termism, Population and the Destruction of the Planet. A Report to the Club of Rome*. New York: Springer.

The Ecological Citizen: www.ecologicalcitizen.net

About the Author

Fred Hageneder is a naturalist who has written eight books on the ethnobotany of trees and on woodland ecology. Some of his work has been translated into nine languages (Spanish, German, Dutch, French, Italian, Polish, Czech, Estonian, and Japanese).

Fred Hageneder is a founding member of the Ancient Yew Group (AYG), an independent research group working to protect ancient yews in the UK. His two monographs on the European yew have met with broad academic approval and generated invitations to appear as a speaker at congresses across the UK, Europe, and as far as Turkey. He works temporarily as external reader (Pacifica Graduate Institute in Carpinteria, California) or peer reviewer (University of Chicago Press) for ethnobotanical topics.

Hageneder is a member of SANASI, a multidisciplinary international group of scientists who since 2013, with the support of the Open University, have been supporting indigenous guardians around the world to protect their natural refuges from land grabbing and destruction. Furthermore, he is a member of the Ecocentric Alliance, a network of university professors, ecologists, conservationists and activists advocating ecocentrism and deep green ethics.

Fred Hageneder lives in Wales as an author, musician, graphic designer and lecturer.

www.healthyplanet.one
www.themeaningoftrees.com

Index

PAGANISM & SHAMANISM

What is Paganism? A religion, a spirituality, an alternative belief system, nature worship? You can find support for all these definitions (and many more) in dictionaries, encyclopaedias, and text books of religion, but subscribe to any one and the truth will evade you. Above all Paganism is a creative pursuit, an encounter with reality, an exploration of meaning and an expression of the soul. Druids, Heathens, Wiccans and others, all contribute their insights and literary riches to the Pagan tradition. Moon Books invites you to begin or to deepen your own encounter, right here, right now. If you have enjoyed this book, why not tell other readers by posting a review on your preferred book site.

Other books in the *Earth Spirit* series

Belonging to the Earth
Nature Spirituality in a Changing World
Julie Brett
978-1-78904-969-5 (Paperback)
978-1-78904-970-1 (e-book)

Confronting the Crisis
Essays and Meditations on Eco-Spirituality
David Sparenberg
978-1-78904-973-2 (Paperback)
978-1-78904-974-9 (e-book)

Eco-Spirituality and Human–Animal Relationships

Through an Ethical and Spiritual Lens

Mark Hawthorne

978-1-78535-248-5 (Paperback)

978-1-78535-249-2 (e-book)

Environmental Gardening

Think Global Act Local

Elen Sentier

978-1-78904-963-3 (Paperback)

978-1-78904-964-0 (e-book)

Honoring the Wild

Reclaiming Witchcraft and Environmental Activism

Irisanya Moon

978-1-78904-961-9 (Paperback)

978-1-78904-962-6 (e-book)

Saving Mother Ocean

We all need to help save the seas!

Steve Andrews

978-1-78904-965-7 (Paperback)

978-1-78904-966-4 (e-book)

The Circle of Life is Broken

An Eco-Spiritual Philosophy of the Climate Crisis

Brendan Myers

978-1-78904-977-0 (Paperback)

978-1-78904-978-7 (e-book)